An Operations Guide to Safety and Environmental Management Systems (SEMS)

An Operations Guide to Safety and Environmental Management Systems (SEMS)

Making Sense of BSEE SEMS Regulations

MICK WILL

Gulf Professional Publishing
An imprint of Elsevier

British Library Cataloguing-in-Publication Data
A catalogue record for this book is available from the British Library

Library of Congress Cataloging-in-Publication Data
A catalog record for this book is available from the Library of Congress

ISBN: 978-0-12-820040-7

For Information on all Gulf Professional Publishing publications
visit our website at https://www.elsevier.com/books-and-journals

Publisher: Joe Hayton
Senior Acquisitions Editor: Katie Hammon
Editorial Project Manager: Xun Wang
Production Project Manager: Anitha Sivaraj
Cover Designer: Greg Harris

Typeset by MPS Limited, Chennai, India

Working together
to grow libraries in
developing countries

www.elsevier.com • www.bookaid.org

Contents

Disclaimer

While the Author has used best efforts in preparing this book, he makes no representations or warranties with respect to the accuracy or completeness of the contents of this book and specifically disclaims any implied warranties of merchantability or fitness for a particular purpose. No warranty may be created or extended by sales representatives or written sales materials. The advice and strategies contained herein may not be suitable for your situation. You should consult with a professional where appropriate. The Author shall not be liable for any loss of profit or any other commercial damages, including but not limited to special, incidental, consequential or other damages.

CHAPTER 1

Purpose and introduction

Contents

1.1 Purpose

This book is intended for the reader who finds themselves working in or associated with offshore oil and gas operations that are covered under the Federal Code of Regulations, Title 30, Part 250 Subpart S, "Safety and Environmental Management Systems" (30 CFR 250 Subpart S). As will be discussed in Chapter 2, "The History of SEMS", across the offshore industry this is also simply referred to as the "SEMS" regulation. The purpose of this book is to satisfy a demand for information regarding this regulation that is not written primarily for regulatory, enforcement or audit personnel. This book is written from an operations perspective.

The information contained herein will not guarantee you or your organization any level of safety or environmental performance. It will not guarantee that your organization will be in full compliance with 30 CFR 250 Subpart S or any other regulation or requirement. Nor will it guarantee that you or your organization will be free from 30 CFR 250 Subpart S enforcement actions or have an audit with no adverse findings. Each organization and operation is different and the specifics of each organization's SEMS program are unique to that organization and its ability to implement.

What it is intended to do is to take what has become a relatively complex topic, and help the reader to understand how all the pieces fit together and to focus on the implementation of a Safety and Environmental Management System that complies with the regulations. My experience has been that organizations and individuals can spend a significant amount of time and resources trying to understand the

An Operations Guide to Safety and Environmental Management Systems (SEMS)
DOI: https://doi.org/10.1016/B978-0-12-820040-7.00001-6

regulations before they can take the first step in developing their program and beginning implementation. I have also seen adverse audit findings caused by organizations simply not understanding the regulations.

As a former operations manager and currently an operations consultant, I am familiar with the difficulty that comes with trying to understand a regulatory requirement and then translate that into actionable objectives. Generally, the reference materials are limited to those that were prepared by and for regulatory specialists or regulatory enforcement organizations. This book is intended to facilitate an operations professional a head start in understanding and implementing the requirements of 30 CFR 250 Subpart S.

1.2 Introduction

Despite the fact that the SEMS requirements became effective November 15, 2010, I am constantly surprised at the variability in the understanding of SEMS across the Gulf of Mexico (GOM) Oil and Gas Operations community. There are many organizations who are working diligently to understand and implement the requirements with varied results. However, to some it is just another regulatory series of hoops to jump through, a periodic pain in the posterior audit process, and to others it is something that the HSSE or regulatory group takes care of.

The confusion is actually understandable, not because the concepts are difficult or new, in fact, operations management systems have been around for a long time. The complexity began in SEMS as a result of how the requirements were introduced and implemented, and complexity remains a part of SEMS as a result of the interworking and relationships between the multiple organizations responsible for the implementation, assurance and enforcement processes.

There is confusion around what it means to have a good SEMS Program. Does it mean I passed the last audit with no findings of non compliance? Does it mean I have a manual covering all 17 Elements of the SEMS requirements? Does it mean my staff can articulate what SEMS is when asked by the auditor? To clarify what good is we need to first understand the original purpose and intent of the requirements, as well as some fundamentals of what a management system is. The good news is that with a little effort and a little time, uncovering the real purpose of SEMS is simple, and the concept of what good looks like in a SEMS Program is relatively clear. We just have to wade through nine years of development and confusion and get back to the basics from which SEMS arose.

Now the next question you should be asking yourself is who is this person telling me they can clarify this whole thing and why should I invest the time to read the rest of this book? Let's get this out of the way right up front. I have never worked for BSEE, I was not involved in the development of their SEMS requirements, and I have never been a part of any regulatory compliance team or organization. Prior to my involvement with SEMS, my background was 32 years of upstream and pipeline operations engineering and management, both onshore and offshore. I was the guy on the side of the table being audited or explaining how I would assure compliance with the regulations that were applicable to my operations.

I tend to think I became involved in the implementation of SEMS on "Day Two" of this now nine year journey. My passion for this subject, while a result of 32 years of accountability to some degree for HSE performance, became a career focus April 21, 2010. On April 20, 2010 at 9:45 p.m., the explosion occurred on the Deepwater Horizon. At that time I was the Gulf of Mexico Region Operations Manager for BP Pipelines North America, and a member of the BP Gulf of Mexico Emergency Response roster. I actually had completed the process for a June 1, 2010 retirement date, having decided I wanted to exit the world of Big Oil and experience either a smaller company or try my hand at consulting before I got too old. While the retirement date remained, the future was going to be way different than I had planned. My friend and colleague, Jim Black, called me the afternoon of the 21st, after he had spent 16 hours in the BP Houston Crisis Center and said "Chief, I need you to spell me so I can get some rest, we are going to be here awhile".

So off I went to the Crisis Center where I had spent many days in the past responding to hurricanes and their destruction, including a stint as Incident Commander for week one of Katrina. What I saw stopped me in my seasoned responder tracks. On the monitors was the still burning Deepwater Horizon, in the break out rooms were people working on how to attempt to activate the BOP with an ROV, and worst of all 11 unaccounted for people. My Deepwater Horizon journey continued through April 2011, and included an assignment as the Deputy Area Commander for the Joint Response headquartered in New Orleans.

After leaving the response organization I accepted a role with BP Gulf of Mexico Operations associated with revising Operating Procedures,

Management of Change and other aspects of their Operations Management System. It was not long before I was eyeball deep in SEMS program development and participated in SEMS Audits as a Subject Matter Expert in 2013 and 2014. In 2015 I completed the required training to be an Audit Team Member auditing GOM Operators' SEMS Programs and I have around 1500 hours as a SEMS auditor, and continue to work as an auditor. Additionally, I have worked with GOM Operators to develop, implement and assess their SEMS Programs.

Why do I tell you all this? Please understand my point of view. I have been involved with SEMS from the beginning, but my roots are in operations. Hopefully I can explain things in such a way that it is clear to those in operations roles who are the frontline of SEMS implementation. I have seen video of the memorial cap on the Macondo well, with the 11 stars representing each of the lives lost in the Deepwater Horizon incident. I have stood in front of the memorial statue in front of Transocean's office. We can never forget what occurred on April 20, 2010.

Suggested reading

Code of Federal Regulations, Title 30, Part 250, Oil and Gas and Sulfur Operations in the Outer Continental Shelf, Subpart S, Safety and Environmental Management Systems (SEMS).

CHAPTER 2

History of SEMS

Contents

Deepwater Horizon well site, Louisiana-hutterstock_779897146 By Breck P. Kent

Before you can really understand the intent and goal of SEMS, you need to understand the history. Otherwise it is easy to get overwhelmed by people talking about SEMS One, SEMS Two, Audit Services Providers, the Center for Offshore Safety, Audit protocol, BOEMRE, BSEE, BOEM, API 75, CFR 250, etc. Like any other aspect of the Oil and Gas industry SEMS has given birth to its own language and acronyms. When someone refers to SEMS, it is important to understand if they are referring to a safety and environmental management system, or are they referring to the regulations in total. In the language of GOM offshore operations. "SEMS" is thrown about in many contexts ranging from the actual management system to the entire spectrum of the regulatory requirements.

Step back to May 2010. There was significant political and public pressure to know how something like the Deepwater Horizon Incident could have occurred. Why was there not sufficient government oversight to protect us from what would become the largest oil spill in U.S. history? The Minerals Management Service (MMS), a division of the Interior Department, was the agency responsible for oversight of U.S. offshore drilling. The MMS had indeed issued the permits for the drilling operations being conducted by the Deepwater Horizon for BP. Investigations were initiated into MMS potential conflicts of interest and potential poor conduct of some employees among other things. Consequently, on May 19, 2010 Secretary of the Interior, Ken Salazar, announced that the MMS would be split into three new federal agencies and temporarily renamed the Bureau of Ocean Energy Management, Regulation and Enforcement (BOEMRE). On October 1, 2011 the BOEMRE was dissolved and replaced by; the Office of Natural Resources Revenue (ONNR), the Bureau of Ocean Energy Management (BOEM), and the Bureau of Safety and Environmental Enforcement (BSEE). There, now you know the alphabet soup of agencies that you may see in the files. Depending on the age of the facility it would not be out of the realm of possibility to see correspondence from MMS, BOEMRE, BSEE, BOEM and ONNR in the same historical files. This has caused confusion and frustration as those unfamiliar with the history begin looking at historical files. From this point forward this book will deal only with BSEE.

Bureau of Safety and Environmental Enforcement Logo.

Can we finally get around to this thing called SEMS? Yes we can, but get ready for another onslaught of acronyms. The original Workplace Safety Rule became effective November 15, 2010. This is often referred to as SEMS I (so you know there is at least a SEMS II coming). Operators on the Outer Continental Shelf (OCS) were required by SEMS I to implement a SEMS Program by November 15, 2011 and submit to BSEE their first completed audit of their SEMS Program by November 15, 2013.

So what is a SEMS? Strictly speaking a SEMS is a safety and environmental management system. Chapter 3 will provide more details regarding management systems and the concepts behind them. For now, let's focus on the SEMS required by BSEE and where it came from. In 1993, the American Petroleum Institute (API) published Recommended Practice Number 75 (RP 75). RP 75 was "Recommended Practices for Development of a Safety and Environmental Management Program for OCS Operations and Facilities". So, adding to the confusion you may see references to a safety and environmental management program (SEMP), but it is pretty much the same as a SEMS. And yes, you read that right, 1993. The Third Edition of RP 75 was published in May 2004 and reaffirmed May 2008. API RP 75 sets out the recipe for a management system comprised of twelve elements (we will get into what elements are in Chapter 3, "Management System Basics"). It also does this in less than 30 pages so there is a lot of room for interpretation of the requirements. Now back to SEMS I. The original Workplace Safety Rule required OCS Operators to develop and implement a SEMS program following API RP 75.

This begs the question, if RP 75 had been around since 1993, SEMS I should have been no big deal, right? Wrong. You see, a Recommended Practice is just that; a recommendation. So, operators were free to implement it, implement parts of it, or just plain ignore it. There are approximately 80 operators in the Gulf of Mexico which means there were 3256 interpretations of RP 75 (yes I made the number up). So off we all went, reading RP 75 and developing SEMS Program documents to be implemented by November 15, 2011. I have talked to individuals with significant experience in offshore operations who purchased a copy of API RP 75 for the first time when SEMS I was rolled out.

Next came that pesky requirement to perform an audit of your SEMS Program implementation by November 13, 2013 and submit it to BSEE. The audit requirements in SEMS I did not provide a lot of guidance regarding this audit. There were questions regarding who should do the audit, could an operator self audit, and what the audit report submitted to

BSEE should look like. Consequently, the first round of audits took on many different processes and looks.

The first round of audits were done and submitted to BSEE. In a memorandum of July 23, 2014 from the BSEE Office of Offshore Regulatory Programs to the BSEE Director, it was concluded that "Based on the first cycle of BSEE audits, the general finding is that the current status of SEMS implementation is geared toward compliance. Operators, in general, did not provide evidence that they are implementing SEMS as an effective management tool". The bottom line is that the first round of audits were not completed in a consistent manner or method, and while it showed that 86% of regulated operators could demonstrate implementation of a SEMS, it was difficult to determine much about the state of implementation from the audits submitted. Some of you may read this and wonder why the audit criteria was not better defined, but it is important to note that while imperfect this was a start. Operators were at least looking at SEMS and implementation to some degree was occurring.

So what happened next? You guessed it SEMS II. Just when you thought it was safe to go back in the conference room (for you youngsters that is a reference to the movie Jaws II). SEMS II became effective June 4, 2013, and operators had until June 4, 2014 to comply with SEMS II. SEMS II increased the SEMS Elements from 12 to 17, and provided additional details regarding some of the expectations of the program. The big impact was the requirement that the second round of audits be completed using an accredited third party Audit Service Provider (ASP). According to CFR 250 Subpart S, after the first audit in 2013 was completed, operators must audit 15% of their operations every three years. Now it is getting interesting. The second round of audits would be coming up in 2015 and 2016. Where do you find an ASP, who accredits them, how will they perform the audits, etc.?

Enter the Center for Offshore Safety (COS). Now let's add a bit of additional confusion at this point. Who is COS, where did it come from and what is their role? In response to recommendations included in the Presidential Oil Spill Commission investigation into the Deepwater Horizon incident (published January 2011), the American Petroleum Institute (API) approved the charter for the COS in March 2011. The API is the primary trade association representing all aspect of the petroleum industry. According to the COS Website, the COS was chartered to operate around the already existing RP 75, which was also the basis for the Workplace Safety Rule of November 15, 2010. With SEMS II, COS was designated as the Accreditation Body (AB) for the Audit Service Providers. The current list of Accredited ASPs at any point in time can be found on the COS Website.

In the ensuing years, COS has produced a variety of items associated with SEMS compliance which are all available on the COS Website. Some of the key documents on the COS Website are:

- COS-1-01 COS SEMS II Audit Protocol-Checklist
- COS-1−04-1 COS SEMS Terms-Definitions Clarification Document
- COS-1−05 COS SEMS Knowledge and Skills Documentation Worksheet
- COS-3-01 Guidelines for Leadership Site Engagement
- COS-1−05 Skills and Knowledge Management System Guideline
- COS-3-03 Guidelines on Maturity Self-Assessment
- COS-3−04 Guidelines for a Robust Safety Culture

It is recommended that when referring to these documents that you go to the COS Website for the most recent version. These are revised and updated and it is best to always make sure you have the most current revision. Additionally, the COS Website has a link to a read only version of API RP 75 that can be accessed free of charge.

Now you have it. The alphabet soup of SEMS that sets out how an operator complies with the requirements and how that operator demonstrates compliance via a third party audit process. Just keep in mind that as you navigate this interrelated complexity of organizations and documents that the original purpose that set this all in motion was very simple; to prevent additional incidents.

Suggested reading

Center for Offshore Safety Website, https://www.centerforoffshoresafety.org/.

Code of Federal Regulations, 30 CFR 250, Oil and Gas and Sulfur Operations in the Outer Continental Shelf, Subpart S, Safety and Environmental Management Systems (SEMS), 7/1/2013 edition.

Griffin, D., Silverlieb, A., Tanneeru, M., Censky, A., May 27, 2010. MMS was Troubled Long Before Oil Spill, http://www.cnn.com/2010/POLITICS/05/27/mms.salazar/index.html#.

Morris, D., Issue memorandum, SEMS Program Summary − First Audit Cycle (2011−2013), https://www.bsee.gov/.

Recommended Practices for Development of a Safety and Environmental Management Program for OCS Operations and Facilities, American Petroleum Institute Recommended Practice 75, May 2008 edition

Reorganization, https://www.bsee.gov/.

Safety and Environmental Management Systems (SEMS) Fact Sheet, https://www.bsee.gov/.

CHAPTER 3

Management system basics

Contents

It is necessary to start with some basics before heading into the world that is Gulf of Mexico "SEMS". As with so many things in the energy industry, there is a relatively simple concept behind what has evolved into a complex process with a life of its own and multiple organizations supporting it. The acronym SEMS stands for safety and environmental management system. The keywords here are "management system". These two words make the BSEE SEMS regulations different from most other regulations "upstream" operators are accustomed to working within. The Oil and Gas industry is often referred to in terms of upstream, midstream and downstream. The SEMS requirements and this book focus primarily on what is commonly referred to as the upstream. I have never found a formal definition of the "upstream", and within specific organizations there may be instances of inconsistency with my definition, but after 40 years in this industry here is my delineation of the three "streams":

- Upstream is all activities associated with the finding, accessing, producing and gathering of oil and gas. This can include exploration activities, production operations and getting the oil and gas to a central point for transportation to market or refinement.
- Midstream is primarily pipelines. Big pipelines. These are the systems that take the oil and gas from producing operations to processing facilities which process the oil and gas into product streams.
- Downstream is gas processing and oil refining. This category may well have the most exceptions resulting from how an organization is organized. In some organizations the gas processing facilities are considered a part of the upstream and in some organizations they are part of the midstream. The pipelines and other transportation facilities that take

An Operations Guide to Safety and Environmental Management Systems (SEMS)
DOI: https://doi.org/10.1016/B978-0-12-820040-7.00003-X
11

the product streams from the gas plants and refineries may be part of the downstream but I also seen them as part of the midstream.

Most upstream regulatory requirements consist of a set of rules or quantitative limits an organization is requires to operate within. If the operations go outside of the prescribed limits, there are consequences. An upstream operation, regardless of onshore or offshore is subject to any number of regulations related to air emissions, loss of containment of fluids, noise limitations, and hazardous waste handling; just to name a few. Excursions outside of the prescribed limits result in anything from monetary penalties to civil and criminal charges. The system is simple. If the rules are followed, nothing happens. If the rules are not followed there are consequences. If oil escapes containment on an offshore facility and gets into the water, there are regulatory consequences. If hazardous waste is not properly disposed of there are regulatory consequences.

When the "rules" are broken, the operator faces the consequences, but what is not generally a part of the regulations are suggested methods to prevent future occurrences. In fact, excessive occurrences can simply result in the operation being shut down. This is what differentiates SEMS. The Deepwater Horizon incident in 2010 demonstrated how significant a major incident in the offshore environment can be. Eleven lives were lost; miles of shoreline contaminated with oil, and clean up costs reaching tens of billions of dollars. SEMS was enacted to prevent future incidents. As a result, SEMS requires the operator to implement a management system intended to prevent future environmental and safety incidents.

This is the heart of how the SEMS regulations are so very different from the rest. SEMS does not mandate a set of rules or limits, but rather mandates implementation of a management system. To illustrate the difference to my clients I use the example of an offshore facility that has recurring releases of oil into the ocean. There will be regulations that are broken by these releases resulting in multiple and probably escalating fines to be paid by the operator. A SEMS auditor or subject matter expert would want to investigate what is failing in the management process that allows recurring releases.

What is a management system? If you do an internet search of "management system", you will get back everything from one sentence descriptions to information on MBA programs with a focus on optimizing the effectiveness of management systems. The International Organization for Standardization (ISO) defines a management system as:

A management system is the way in which an organization manages the inter-related parts of its business in order to achieve its objectives. These objectives can relate to a number of different topics, including product or service quality, operational efficiency, environmental performance, health and safety in the workplace and many more.

One of the definitions I like the best is from Wikipedia (really):

A management system is a set of policies, processes and procedures used by an organization to ensure that it can fulfill the tasks required to achieve its objectives. These objectives cover many aspects of the organization's operations (including financial success, safe operation, product quality, client relationships, legislative and regulatory conformance and worker management). For instance, an environmental management system enables organizations to improve their environmental performance and an occupational health and safety management system (OHSMS) enables an organization to control its occupational health and safety risks, etc.

Many parts of the management system are common to a range of objectives, but others may be more specific.

A simplification of the main aspects of a management system is the 4-element "Plan, Do, Check, Act" approach. A complete management system covers every aspect of management and focuses on supporting the performance management to achieve the objectives. The management system should be subject to continuous improvement as the organization learns.

Elements may include:
- *Leadership Involvement & Responsibility*
- *Identification & Compliance with Legislation & Industry Standards*
- *Employee Selection, Placement & Competency Assurance*
- *Workforce Involvement*
- *Communication with Stakeholders (others peripherally impacted by operations)*
- *Identification & Assessment of potential failures & other hazards*
- *Documentation, Records & Knowledge Management*
- *Documented Procedures*
- *Project Monitoring, Status and Handover*
- *Management of Interfaces*
- *Standards & Practices*
- *Management of Change & Project Management*
- *Operational Readiness & Start-up*
- *Emergency Preparedness*
- *Inspection & Maintenance of facilities*
- *Management of Critical systems*
- *Work Control, Permit to Work & Task Risk Management*

- *Contractor/Vendor Selection & Management*
- *Incident Reporting & Investigation*
- *Audit, Assurance and Management System review & Intervention*

What does a management system look like? There are some clues in the above definitions that can be used to form a very basic high level understanding of management systems that is sufficient for the understanding of SEMS. When I am asked to explain the concept of a management system to a client or during an audit, I focus on five high level things that an auditor looks for when judging the effectiveness of a SEMS program:

- Elements
- Policies, Processes and Procedures
- Objectives
- Verification
- Continuous Improvement

In management system terminology a management system is broken into what are called "elements". So what is an element, and what makes a good element? Many operators in the energy industry will have some type of high level statement that is along the lines of no harm to people and no harm to the environment. While that sounds like a great goal, think about all the things that must be done correctly and consistently to make such a goal a reality. People must wear the correct PPE, they must follow safe procedures, they must report unsafe conditions, they must recognize hazards, they must manage change, etc. The list goes on and on. It is simply not possible to be able to effectively manage this as a single process or operation. How do you eat a whale — one bite at a time. I look at elements as bite size pieces of an operation that can be effectively managed. By effectively managed I mean that objectives can be developed, processes defined and improvement measures monitored. For example, consider the issue of personnel wearing the correct PPE. We can set a goal of 100% compliance, meaning everyone wears the correct PPE every day. We can define a process for educating the staff regarding the requirement and develop a checklist or observation program to monitor performance. If the monitoring shows less than 100% compliance then we can develop an action plan to modify the process to affect improved performance.

There is a potential resource wasting, headache causing aspect of elements that can be avoided. I have observed organizations spend countless hours on how many elements their system should have and how to identify or name them. Wikipedia says that a management system can include up to 20 elements. I personally have seen management systems with eight, fourteen, and seventeen elements. The true answer is it just does not matter as long as the system works for your organization. Additionally, if you find your system is not working as well as you would like, you can change the number of elements, this is not a "forever" decision. It is better to take a first shot at the elements, implement your management system and optimize with real world experiences. Allow me to depart from the general discussion of management systems and address the SEMS requirements specifically. The Bureau of Safety and Environmental Enforcement has chosen to characterize an efficient Safety and Environmental Management System via 17 elements. Later chapters of this book will delve into these 17 elements in detail, but for now just make a mental note regarding 17. Consequently, if you develop a SEMS program using 10 elements, you will have introduced the potential for complicating the audit process which we will discuss beginning in Chapter 6, "Getting SEMS Started".

Moving to the "three P's" policies, processes and procedures. The more linear thinkers among us will want a definition of each of these, and I have to confess to having spent some hours working on this very quandary in my years as an operations manager. After finally reaching a definition for each of them that the organization could agree upon and accept, then the document library had to be sufficiently controlled to prevent policies from getting put in with the procedures and never let the processes be mistaken for a policy. As an auditor, I would tell you it just does not matter. What truly matters is that within the elements there are documents that describe "how" things are to be done. From an auditor perspective, if it is not documented it does not exist. Before you let this comment send you off working out exceptions to it, let me clarify. The complexity of the operations will determine the amount and granularity of the required documentation. Most offices have a one page document that fully describes the management system for making coffee in the lunchroom. Depending on the office coffee maker, this may include controls on the measuring of the coffee and the required disposal of grounds and filters. But without the documentation, it is very unlikely that everybody will make the coffee in the same way, which makes the quality of the coffee variable. We will not get into the person who ignores the management system, which is a personal performance issue not a management system issue. Conversely, the management system documents associated with starting up an offshore facility after an emergency shutdown will be significantly larger than those for making coffee.

Consequently the first thing an auditor will likely ask you for are the documents that describe how the work within any element is to be accomplished. This is because consistency in operations is just about impossible without documented guidance. Operating procedures are an obvious example. Say an organization determines that the desired method to start up a compressor on a facility follows a number of steps to be completed in a specific order, and that this method has been determined to eliminate risk to the personnel involved and risk of damage to the equipment. Now let's assume the operations personnel work on 14 day rotations, and there is a day and night shift within each rotation. This means that there are a lot of people who may potentially be involved with starting that compressor. The only way to assure consistency is to have the process documented (and that still does not guarantee 100% consistency).

Having the documentation complete is just the start. This may sound a bit elementary, but the document has to be correct. The big question is what is correct with respect to a procedure or policy? The answer is two parts. The first is that technically it must result in the desired outcome; whatever that is. For example, a start up procedure for a glycol dehydration unit must result in the unit being started up without risk to personnel, risk of loss of containment of glycol, or risk of damage to the equipment. The procedure is likely subjected to technical reviews, operational reviews, peer reviews and risk assessments all of which are outside the scope of this discussion. Just remember, it has to be technically correct.

The second part of the answer is the one that this discussion needs to consider. When there are regulations associated with the operation, the policies, procedures and programs must comply with the requirements as mandated by the regulations. For example, if there is a regulatory requirement that operating procedures are to be reviewed every 5 years, your document needs to be consistent with this. Having a documented process that requires review every 15 years does not meet the regulations and your process is not acceptable. Again, this may sound elementary, but in large organizations there can be requirements set in corporate offices with no knowledge of localized regulations. An organization may have a document retention policy that requires certain documents to be maintained for 3 years, but the local regulations require 6 years. While you can satisfy corporate requirements with 3 years you would not be in compliance with the regulations under which you operate. I have witnessed operators plead their case in great detail and with great conviction that their internal requirements have been developed with great care and that compliance with them should be sufficient to satisfy the regulatory requirements. The regulators and auditors then reply with some language explaining that the regulations don't work that way and that it is still a finding of non-compliance.

It is important that your elements, whatever you decide to name them, are a vehicle for improvement. This means that there should be some objectives associated with each element. I explain this using an analogy to what are the elements of a healthy lifestyle. We are told that for overall health we should eat a nutritionally correct diet, get regular exercise, get sufficient sleep, stop using tobacco, moderate our intake of alcohol, have regular checkups, etc. It seems the list gets longer every day, but that is not the point here. If we make improvements in any one of these items, we can impact our overall health. The same is true of the elements of an effective management system. While multiple elements are required for an effective management system, making improvements in specific elements should impact the quality of the overall operation. A common management system element is hazard analysis. By improving the hazard analysis process, more potential hazards can be identified and mitigated. This can result in less impact to personnel, the environment and generally reduces downtime.

After the audit team has confirmed that you indeed have a documented management system and that the management system as documented meets the requirements for compliance, the next area of focus is the evidence that the organization actually follows the management system. As an auditor, you get an immediate feeling regarding how the rest of the audit is going to proceed at this stage. Those organizations that have a robust internal verification process can transition with little effort or hesitation into the discussion regarding how well their organization follows their management system. This does not mean that every organization with a verification process is fully implementing their management system, but rather that they are aware of where they need to improve and where they are excelling. On the other hand, it is very obvious when the operator's personnel do not know if they are following their management system. The tension in the room grows, people fidget in their chairs, they look to others to speak, and many times this begins the long process of more and more people being brought into the discussion as the operator looks for someone who knows the status of the management system implementation.

Just because your management system says how something is to be done does not mean it is happening, nor does it relieve the operator of their obligation to make sure it is happening. When the regulator representative finds improper handling of hazardous waste by a contractor on your facility, they do not care that you have a document requiring the contractor to handle the waste in a correct manner. All they care about is that it is not getting done correctly and the operator is responsible. I see this a lot when dealing with elements regarding management of change. I have audited operators who have well documented management of change procedures and extensive software tools for documenting the required steps, and yet review of individual management of change events uncovers that the documented process for managing change is not being followed.

This opens up another aspect of management systems that can lead to hours of discussion around audit tables. The processes, policies and procedures that are documented within a management system must be followed 100% of the time. The scenario I see played out time after time is also best illustrated using management of change. As an auditor I have seen changes made to a facility without following the operator's management of change process. Usually the process was not followed because an individual within the organization made the determination that the change had no significant impact on the facility. I have seen cases where the individual who made that specific decision was technically competent and the change did indeed have no impact on the facility. So what is the big deal? This is a slippery slope that an operator cannot allow themselves to go down. There are so many "what ifs" that come into play. What if the next time this individual thinks they are qualified to make such a decision but in reality they are not? What if other personnel who are not as qualified begin to make such decision on their own? What if several of these changes have a cumulative impact? The only way to minimize risk is to follow the documented process every time, even when it seems like an unnecessary use of time and resources.

My personal view of management systems may well diverge from the definitions seen above in one component. Management system definitions talk a lot about achieving objectives and efficiency improvement. With respect to operations in a potentially hazardous industry such as offshore oil and gas production, I think of control of operations as an objective.

Many times people think of objectives in terms of financials, metrics, statistics, etc. These are quantitative objectives. However, management systems are a good tool to control "how" things are done inside an operation. There is an argument that this control is in support of the quantitative objectives, but I prefer to think of control as an independent objective, especially when we begin to focus on environmental and safety objectives. For example, an organization may have a goal of zero injuries. That is a quantitative objective, but it can be exceeded by one incident. To prevent that one incident requires a significant amount of control with respect to how personnel approach and complete tasks.

Continuous improvement is an overused and over complicated concept. I have seen dozens of continuous improvement forms. I have seen organizations with continuous improvement coaches. Don't get me wrong, if these are working for your organization keep going. I prefer to keep one question in mind, and sometimes I post it in a client's office. Are you doing anything different than you did in the past as a result of your management system? And I suggest you dig deep, go all the way to the "wrench end" where the field operations take place daily. I have observed organizations with well thought out and well documented management systems but over several years of implementation the actual field operations look pretty much the same as they did before all the effort went into the management system development.

There are some key points to remember as this book progresses into more and more detail around "SEMS". At the heart of it all is the management system, which is simply a collection of documents that set forth "how" an organization is going to do things. And how they expect personnel to do things every time. As you become more exposed to the world of oil and gas operations on the US outer continental shelf you will find that when someone talks about SEMS they may well be referring to the entirety of the BSEE regulations which includes the management system, enforcement, inspections, data submission, etc. all of which will be covered in this book. But at the core is a management system.

Suggested reading

Management System Standards, International Organization for Standardization, https://www.iso.org/management-system-standards.html.
Wikipedia, Management system, https://en.wikipedia.org/wiki/Management_system.

CHAPTER 4

SEMS and ISO 14001

Contents

Since the day that SEMS I became effective via the original Workplace Safety Rule on November 15, 2010, people have claimed that BSEE SEMS is just like ISO 14001. Many organizations that were ISO 14001 compliant thought that becoming SEMS compliant would be a "cut and paste" exercise. I was working within an organization where that was indeed the initial opinion. However, it was an opinion formed without looking into the details of both BSEE SEMS and ISO 14001. What we found as we began the journey to SEMS implementation was that while BSEE SEMS and ISO 14001 are related, they are significantly different. As an operations manager I was responsible for an organization that was ISO 14001 certified, and once I began digging into the details of SEMS I the differences became apparent and the true scope of SEMS I implementation began to clarify.

To fully understand this topic, you need to have a basic understanding of who ISO is and how the certification process for ISO 14001 works. "ISO" stands for the International Organization for Standardization. According to their website:

> The ISO story began in 1946 when delegates from 25 countries met at the Institute of Civil Engineers in London and decided to create a new international organization 'to facilitate the international coordination and unification of industrial standards'. On 23 February 1947 the new organization, ISO, officially began operations.

The important thing to note here is that ISO has been around a relatively long time, and is a respected organization. Their scope of influence is also broad, according to their website their organization is made up of members from the national standards bodies of 164 countries and they have published over 22571 International Standards. Of these 22571

An Operations Guide to Safety and Environmental Management Systems (SEMS)
DOI: https://doi.org/10.1016/B978-0-12-820040-7.00004-1
21

International Standards we are going to focus on only one of them; ISO 14001.

ISO 14001 was first published in 1996, and the focus of ISO 14001 is the implementation of an environmental management system with the goal to reduce an organization's environmental impacts. Similar to SEMS, ISO 14001 utilizes a management system to accomplish the objectives. As with most management systems, ISO 14001 follows the Plan–Do–Check–Act model. ISO 14001 provides a roadmap to implementation of an Environmental Management System (EMS), with the goal of this implementation to minimize an organization's environmental impacts. ISO 14001 does not contain specific goals or requirements, but rather how to implement a system to significantly reduce environmental impacts. Based upon my personal experiences, a key word here is "significant". An organization is deemed ISO 14001 compliant via an assessment by an accredited ISO 14001 audit team. While ISO 14001 does not contain specific goals or requirements, the assessment process does look to the implementation of the EMS such that it addresses areas where the impacts represent significant improvement for the organization. Take a pipeline organization for example. If the organization has a history of contamination of the environment due to leaks but chooses to focus their ISO 14001 activities on the office recycling program and not the reduction of leaks, they are not implementing in the spirit of ISO 14001 and not likely to be seen to be ISO 14001 compliant.

What follows should in no way be construed as reference material for ISO 14001 content or certification. This is simply enough information regarding ISO 14001 such that the differences between SEMS and ISO 14001 can be discussed. As a management system, ISO 14001 has six primary "clauses" or elements:

1. General Requirements
2. Environmental Policy
3. Planning
4. Implementation and Operation
5. Checking and Corrective Actions
6. Management Review

As discussed in Chapter 3, "Management System Basics", the number of elements is not particularly indicative of the process thoroughness. So the fact that ISO 14001 has six elements and SEMS has 17 elements is not significant to this discussion. What is important is what are the expectations associated with each of these six areas. By following an organization's progression through these areas the ISO 14001 process is relatively easy to understand at the level appropriate for the purpose of this book.

With respect to the comparison between ISO 14001 and SEMS, I think the largest point of commonality is the "General Requirements". If an organization intends to implement any kind of management system, there is a foundation that must be developed or the discipline required to implement the management system will suffer. As an individual who has been involved with the implementation of both ISO 14001 and SEMS within an organization, the ISO 14001 General Requirements and the SEMS General Element both require:

- Support of management
- A documented policy focused on performance improvement
- An appropriate review of associated metrics
- Verification of the desired activities

This may look rather elementary to some readers, but an organization must fully realize what they are pursuing, be committed to pursuing it, and make it clear to everyone in the organization that this is a desired path. Being compliant requires effort and resources. These resources are likely being pulled in multiple directions with multiple priorities. Without the organizational commitment to compliance there is no assurance that the compliance effort will get the resources and attention required. If compliance is seen as the role of the regulatory team or the HSE team, its importance to the rest of the organization is diminished significantly.

The rest of the ISO 14001 process is the implementation of the plan-do-check-act model, which per Chapter 3, "Management System Basics" is behind any type of management system. The next step for an organization which is dedicated to reducing its environmental impact is to identify what that environmental impact is. While there are many methods for accomplishing this identification, the organization is left to determine the best method for their operations. Just remember that when assessment time comes the organization will be required to describe how the impacts were determined. In my experience, identification of the environmental impacts is the easier part of what is a two step process. Step two is to prioritize these impacts with respect to where the biggest improvement can be realized. I have spent hours in conference rooms comparing big projects that take a long time but yield big results to projects that yield smaller results but can be done quickly and make an immediate impact. Using the pipeline operation as an example again, the best way to prevent an aging system from leaking may be to replace it with new pipe. Depending on the size of the project, it may take years before a reduction in leaks is realized. Conversely, the organization may be able to utilize new inspection technology to try to identify the highest risk sections and address those immediately. While this can have a much quicker impact, it may not have the degree of impact a full replacement would.

I am sure many of you are already thinking about the next step, because it is the same for any management system. Once the decision is made regarding what to do comes the stage where we determine who is going to do it, how long it will take and how much it will cost. With respect to the ISO 14001 process, the activities, methods and tools an organization employs to achieve this reduction in environmental impact are left largely to the organization to specify. Another reminder; once an organization specifies how the results are going to be achieved the methods are subject to review in the assessment phase. Consequently, different organizations in different industries will likely be utilizing different methods and tools. For some the answer may be implementation of procedures, for another it may be development of a training program, for another it might involve an assessment of the organization's physical infrastructure.

I am now going to skip a lot of details that are near and dear to those who are heavily involved in the implementation of ISO 14001, not because they are not important but because they are not relevant to this discussion. Once an organization has implemented the activities identified in the preceding process, "plan and do" are complete and "check" should be in full operation. Simply put, check is an assurance that the desired results of the activities are occurring. Many of us have been involved with projects or activities that were implemented with the best intentions and rigor, but the results just did not follow.

This is where "act" takes center stage. With respect to ISO 14001 compliance, if the desired results are being achieved, it is time for the organization to act to move on to the next project or activity on the priority list. If the results are not meeting expectations, the organization should assess if changes are needed and act to implement the changes. In my experience this is where the assessment process shifts to a focus on continuous improvement. It is one thing to get that first assessment that indicates your organization is ISO 14001 compliant, but to maintain that status the program must continue to progress. This is not a onetime "certification".

What makes SEMS different? First off, SEMS is specific to one sector of one industry; the offshore oil and gas operations on the US Outer Continental Shelf. This specificity will show up throughout this discussion. Second, SEMS is a safety and environmental management system as opposed to an environmental management system. To understand this, keep in mind that offshore oil and gas facilities have some unique aspects that can link safety and environmental risks. This is not to say that safety and environmental considerations are not closely linked in other industries but simply that the link in the offshore oil and gas sector is significant.

First, oil and gas operations by nature involve flammable fluids. The whole goal is to produce oil and gas, resulting in a facility that purposely stores and processes flammable fluids. When there is a loss of containment of a flammable fluid, there are both safety and environmental risks associated. Second, offshore oil and gas facilities have limited space. Consequently, these flammable fluids and the personnel associated with the operations of the facility are generally in relatively close proximity. When there is a loss of containment the potential for impact on personnel is relatively high. Third, responding to a loss of containment or a safety incident is logistically complicated and involves its own inherent risks. Response can involve helicopters and boats working offshore, as well as potential for ignition of flammable fluids or contact with potentially harmful substances.

Just to be clear, there are injuries in the offshore oil and gas industry that do not involve environmental impacts, and there are environmental incidents that do not involve injuries. However, especially when considering significant incidents, there is a higher likelihood of an incident involving both safety and environmental impacts than in many other industries.

Since SEMS is limited to offshore oil and gas operations, the requirements can be much more prescriptive than those of ISO 14001. For example, SEMS requires that an operator has a management of change (MOC) process in place and that their management of change process contains specific characteristics. To be in compliance your MOC process must meet these specific requirements. Additionally, since SEMS is limited to offshore oil and gas operations on the US Outer Continental Shelf, compliance with SEMS is not optional but rather a requirement with enforcement responsibility resting with BSEE.

This point warrants further emphasis. Compliance with ISO 14001 is voluntary. I have been a part of many conversations regarding the worth of ISO 14001 compliance to oil and gas production and pipeline operations (I cannot comment on the downstream viewpoint). The conclusions of those conversations are not relevant, but what is relevant is that with respect to a voluntary program organizations will attempt to associate a value or worth to expending the resources necessary for compliance. With respect to compliance with SEMS on the US Outer Continental Shelf for oil and gas operations the exercise of associating a worth to SEMS compliance is a non value adding effort. Compliance and compliance with the specific details of SEMS is a requirement.

Since ISO 14001 has been around longer than BSEE SEMS, many organizations who are familiar to some degree with ISO 14001 find themselves implementing BSEE SEMS. While moving forward with the implementation of a BSEE SEMS program, keep in mind the primary differences:

- BSEE SEMS is focused on both environmental and safety risk reduction
- BSEE SEMS only applies to offshore oil and gas operations in the OCS
- BSEE SEMS is not optional and BSEE has enforcement options
- BSEE SEMS is much more prescriptive than ISO 14001

Suggested reading

International Organization for Standards, 2015. Environmental Management Systems — Requirements With Guidance For Use. International Organization for Standards.

Whitelaw, K., 2004. ISO 14001 Environmental Systems Handbook, second ed., Burlington: Elsevier

CHAPTER 5

SEMS and process safety

Contents

Once you are heading into the world of BSEE SEMS, it is inevitable that someone will tell you that SEMS is just process safety. In my opinion, the relationship between process safety and SEMS is significantly more complicated than that statement portrays. To understand this relationship requires a basic understanding of where process safety came from and what process safety looks like in its purest form. Many of us have some kind of understanding of process safety that goes back to a classroom or a training session, but do we really understand what process safety is and how it applies to our industry?

Regretfully, my process safety awareness was associated with a significant incident. On March 23, 2005 I was at my desk in my role as Gulf of Mexico District Operations Manager for BP Pipelines North America in the Houston Westlake complex. March 23, 2005 was my Daughter's eleventh birthday so I may have been thinking about whatever the plans were for the celebration. Possibly I was getting myself mentally prepared for the sleepover that would be part of the celebration. At approximately 1:20 p.m., there was an explosion heard for miles, and smoke was soon filling the sky in the direction of Texas City. I know it took some time, but it seemed almost instantaneous that the word spread through the office that the explosion had occurred at BPs Texas City Refinery. And it was not long before we heard that there were fatalities. In fact 15 individuals perished in the incident, and more than 170 were injured.

Soon there were multiple investigations underway into the cause of the tragic incident. In addition to the internal BP investigation the U.S. Chemical Safety and Hazard Investigation Board (CSB) initiated an investigation which led to the BP U.S. Refineries Independent Safety Review Panel led by former Secretary of State James A Baker III. This is commonly

An Operations Guide to Safety and Environmental Management Systems (SEMS)
DOI: https://doi.org/10.1016/B978-0-12-820040-7.00005-3

referred to as "The Baker Report", published January 2007. A common theme in all the investigations was the lack of process safety understanding and implementation. Soon the industry was buzzing with definitions and opinions regarding process vs. personal safety. As an operations leader the future contained a myriad of meetings, training, reviews of the Baker Report, and ultimately attending the BP Operations Academy at MIT. This does not make me any kind of authority on the topic, but I did experience the shift in the upstream and midstream sectors of the industry to where process safety became a more common topic of discussion and interest.

What is process safety? According to the American Institute of Chemical Engineers (AIChE) Center for Chemical Process Safety (CPS):

"Process Safety is a disciplined framework for managing the integrity of operating systems and processes handling hazardous substances by applying good design principles, engineering, and operating practices. It deals with the prevention and control of incidents that have the potential to release hazardous materials or energy. Such incidents can cause toxic effects, fire, or explosion and could ultimately result in serious injuries, property damage, lost production, and environmental impact".

After first reading this, it is relatively easy to jump to the conclusion that by implementing SEMS you are managing process safety. I am personally hesitant to make such a direct linkage between SEMS and process safety. This discussion takes us into what is commonly referred to as Process Safety Management (PSM). In "A Brief History of Process Safety Management", Ian Sutton discusses Process Safety Management (PSM) as being composed of the following three core features:

1. *A process plant is a complex system that must be managed holistically. The behavior of individual workers is important, but it does not address systems issues.*

2. *PSM is performance-based—the only measure of success is success.*

3. *It is non-prescriptive. Rather than creating a large number of rules that must be followed, managers determine what they need to do at their facility to ensure safe operations.*

What I like about this description is the emphasis on the understanding of the operations as a system, and the need for the system to be managed as a system. Many of the elements of SEMS are essential to effective process safety management. It would be difficult to envision a PSM that did not require (among other elements) the effective use of:

- Hazard Analysis
- Management of Change

- Safe Work Practices
- Mechanical Integrity
- Training
- Operating Procedures

However, using Ian Sutton's components of PSM, effective implementation of process safety requires a depth of understanding of the system being operated that goes beyond the basic implementation of a SEMS program. The upside is that SEMS implementation which includes this depth of understanding can result in effective process safety management. Explained in another way, it is my opinion that an organization can implement a SEMS program and not implement an thorough PSM program, but an organization with an effective PSM program will indeed satisfy many elements of the SEMS requirements.

Using Management of Change (MOC) as an illustration may help understand this rather vague distinction. Per 30 CFR 250 Subpart S, the requirements of a SEMS MOC process are:

(a) You must develop and implement written management of change procedures for modifications associated with the following:

(1) Equipment,

(2) Operating procedures,

(3) Personnel changes (including contractors),

(4) Materials, and

(5) Operating conditions.

(b) Management of change procedures do not apply to situations involving replacement in kind (such as, replacement of one component by another component with the same performance capabilities).

(c) You must review all changes prior to their implementation.

(d) The following items must be included in your management of change procedures:

(1) The technical basis for the change;

(2) Impact of the change on safety, health, and the coastal and marine environments;

(3) Necessary time period to implement the change; and

(4) Management approval procedures for the change.

(e) Employees, including contractors whose job tasks will be affected by a change in the operation, must be informed of, and trained in, the change prior to startup of the process or affected part of the operation; and

(f) If a management of change results in a change in the operating procedures of your SEMS program, such changes must be documented and dated.

While this definition looks thorough, what is missing with respect to effective management of process safety cannot easily be put into the wording of an MOC definition. For example, an MOC program that involves the deeper, holistic understanding of the process as a whole might include (among other things):

- Requirements for the qualifications and expertise of the technical reviewers
- Requirements for independent third party expert review
- Requirements for the review of the MOC across multiple disciplines and operating areas
- Requirements for the type of hazard analysis to be completed prior to authorization of the change
- Requirement for the delegation of authority to approve the change based upon the hazard analysis findings

While this is potentially a subtle distinction, the key takeaway is that implementing a SEMS program does not necessarily equate to implementing process safety management.

Suggested reading

Baker, J., et al., 2007. The Report of the BP U.S. Refineries Independent Safety Review Panel.

Code of Federal Regulations, 30 CFR 250, Oil and Gas and Sulfur Operations in the Outer Continental Shelf, Subpart S, Safety and Environmental Management Systems (SEMS), 7/1/2013 edition.

Sutton, I., 2018. A Brief History of Process Safety Management, second edition Process Safety FAQs, American Institute of Chemical Engineers. Available from: https://www.aiche.org/ccps/about/process-safety-faqs.

CHAPTER 6

Getting SEMS started

Contents

Armed with a bit of history and understanding it is now possible for you to distill the whole world of SEMS, BSEE, COS, etc. into the high priority items that an OCS operator must do to comply with CFR 250 Subpart S. Now this is not by any means a full list, but it is intended to get you rolling in the right direction and avoid the most common potholes along the way. After over eight years of SEMS implementation in the GOM we can learn from past experiences and best practices. While it is very tempting to dive off into the 17 elements and the details of the regulations. However, it has been my observation that those organizations who have implemented the most effective SEMS programs have taken time to first consider two high level questions:

1. What is the overarching purpose of my SEMS program?
2. How will my SEMS program be assessed?
   When I ask people these questions, there are four basic responses:
1. Some type of answer that involves being in full compliance with all requirements to operate in the GOM
2. Some type of answer that involves wording to the effect of no harm to people or the environment
3. The blank look similar to "a cow seeing the new cattle guard"
4. A discussion regarding risk analysis, internal verification and external auditing.

An Operations Guide to Safety and Environmental Management Systems (SEMS)
DOI: https://doi.org/10.1016/B978-0-12-820040-7.00006-5

6.1 The goal of a SEMS program

Let's start with what CFR 250 Subpart S says the goal of a SEMS program should be:

250.1901 What is the goal of my SEMS program?

The goal of your SEMS program is to promote safety and environmental protection by ensuring all personnel aboard a facility are complying with the policies and procedures identified in your SEMS.

a. To accomplish this goal, you must ensure that your SEMS program identifies, addresses, and manages safety, environmental hazards, and impacts during the design, construction, startup, operation (including, but not limited to, drilling and decommissioning), inspection, and maintenance of all new and existing facilities, including mobile offshore drilling units (MODUs) when attached to the seabed and Department of the Interior (DOI) regulated pipelines.

b. All personnel involved with your SEMS program must be trained to have the skills and knowledge to perform their assigned duties.

Notice that there is no mention of the SEMS audit. The goal of SEMS is not simply to "pass" the periodic ASP audit and to report progress on your actions to BSEE. The goal is to reduce releases to the environment and to increase worker safety. It is that simple, yet it is that complex. Reducing slips, trips and falls is not the same as preventing an uncontrolled release into the GOM. Chapter 5 SEMS and Process Safety, is a discussion of the basics of process safety and how an effective SEMS program does and does not correlate with Process Safety Management (PSM).

To really understand the goal of a SEMS program, it helps to remember what drove the development of SEMS; the Deepwater Horizon incident. If you remember this, the vagueness of some of the wording and questions related to SEMS is not so confusing and leads to the "right" thing to do. Many years ago I attended a law school commencement and one of the speakers was an elderly alumnus of significant accomplishments. He advised the newly minted attorneys to always search for the "right" thing to do. I see SEMS in much the same light. If an organization remembers that the purpose of a SEMS program is to "promote safety and environmental protection" then determining the "right" thing to do is easier.

So far the concept is seems pretty simple, and promoting safety and protecting the environment are great concepts to aspire to. The next question is where this gets difficult; how does this get done? This cannot be solved by making sure personnel all wear the appropriate personal protective equipment (PPE), or by making sure all the stair handrails are solid and the steps marked in the required yellow tape. Incidents that involve serious injuries and releases of material into the environment can be complicated in nature. Many times they involve multiple things failing or going wrong at the same time. Reducing the risk of these types of incidents is as complicated as the incidents are.

Reducing the risk of incidents requires understanding your risks and implementing a system of process that works to reduce these risks. The key word is implementing. A process is not implemented by simply documenting it. While documentation is the first step, the real work begins after the documentation. Documentation is necessary for consistency across operations. The offshore world is complex and always changing. Personnel work "hitches", which consist of a period of time on the facility followed by a similar period of time off the facility. While on the facility, work continues around the clock, so there are multiple shifts during each hitch. Consequently, an offshore worker will work for a variety of leaders on a variety of shifts. Documented procedures are the only way to ensure that each leader has each shift of each hitch perform work processes in the way the organization has determined to be most effective. For leaders to consistently follow a system of processes, it must be viewed as a priority. What makes it a priority? It becomes a priority when the leadership consistently communicates the importance of it to the workforce, from the very top levels to the first line team leaders.

I have been asked by auditors and regulators "what is the purpose of your SEMS program", or "what does SEMS mean to you". I have found that the best answer to these questions is that it all comes down to understanding and eliminating risk.

6.2 Assessing a SEMS program

Here is where operations can take a clue from the auditors. Auditor training focuses on three key questions:
1. Is the process to be followed documented?
2. Does the documented process meet the requirements of the appropriate regulations or legislation?
3. Can the organization being audited demonstrate implementation of and adherence to the documented process?

It is that simple. Three yes answers would tell an audit team (or internal assessment team) that there is a pretty good chance the program is being followed. And if the program were developed to achieve the goal of the SEMS program to "promote safety and environmental protection" then there is a pretty good chance that there is some degree of risk identification and mitigation ongoing in the organization. As discussed in Chapter 6, SEMS and Process Safety, the level of understanding and the application of a holistic understanding of the process being evaluated will determine the thoroughness of the risk mitigation.

Just as simple is the concept that if you are doing everything the same way you always have then you may not be improving. A worthwhile exercise is to do a self assessment of your operations as they were in 2010 and as they are currently. Be honest. I have witnessed operators investing significant effort and resources in attempting to demonstrate how existing processes meet the SEMS requirements despite the processes missing critical components of the requirements. That effort would have been better spent in developing a plan to modify the existing processes to include all the requirements. Have you really implemented changes intended to reduce risk, or have you managed to get through your SEMS audits and subsequent CAPs without really making any changes? If it is the latter you are operating with the same risk profile as you had in 2010, and you should not be comfortable with that in light of what history has taught us. The absence of historical incidents associated with a facility is not an indication of the absence of risk.

Many operators have internal Operations Management Systems (OMS) in place that control their work processes. It is important that in these cases the operator's OMS does indeed address all 17 Elements included in the BSEE SEMS II requirements. As discussed in Chapter 3, Management System Basics, a management system can have any number of elements. Consequently, an OMS can be structured any number of ways and can have the number of elements and levels that the operator chooses. But the OMS must address all 17 SEMS Elements. You may not agree with all 17 Elements, you may not agree with the organization of the 17 Elements, you may not agree with a lot of things about the 17 Elements. It does not matter. You will have to be able to demonstrate compliance with all 17 SEMS II Elements.

"Demonstrate" is quite a word. It has certainly evolved over my time in the industry. The best way to illustrate what I mean is to use examples. In the area of training and competency, there was a time when if the

OIM or the foreman said someone was competent then they were competent. Enough said. This evolved into the OIM or Foreman observing individuals and making written notes confirming their competency. In today's world, notes in the OIMs notebook are just not going to suffice. Today it requires training records, competency assessments, etc. to be documented. Similarly, something like making a change to an Operating Procedure requires documentation (usually via Management of Change) that, among other MOC requirements, personnel impacted by the change were indeed informed of the change and attended additional training if appropriate. Without sufficient documented evidence illustrating that you follow your processes, it is difficult to demonstrate compliance. Remembering the three auditor questions, a nice set of documents on a shelf can at best result in two yes answers out of the three. The processes must be implemented to have three yes answers.

The audit process is becoming more defined all the time, and with the requirement of the participation of an ASP the formality is increasing. By formality I mean that the COS protocol will be used, the audit plan will be reviewed by BSEE, a report will be prepared by the ASP, the operator will develop and submit a Corrective Action Plan (CAP) to BSEE, and continuous improvement will be looked for. The purpose of the audit is twofold. First, it is an opportunity for the operator to demonstrate how they comply with the requirements. During this process an operator can also get input regarding best practices to enhance their program. Second, it is an opportunity to identify those areas where the operator falls short of compliance and needs to improve. I did not mention the words penalty, fine, failure, or any of the others sometimes associated with regulatory issues. Learn from the findings and use them to reduce your risks. Listen to this one point even if you forget the rest of this chapter. Make the audit useful to you and don't unnecessarily complicate the process. The audit process and how to prepare for an audit are discussed in detail in Chapter 8, The SEMS Audit, but some basic recommendations for an effective audit are:

1. The audit requirements are outlined in detail in CFR 250 Subpart S. Follow them. Get your plans submitted, your ASP selected and your arrangements made in time. You are likely on an every three year schedule so there is no reason to miss your deadlines.
2. The audit will utilize the COS protocol. Be familiar with it. My experience as an auditor is that if the first time the audit participants see the protocol is the first day of the audit it will be a long and arduous audit.

3. Be an educated consumer. As the operator you have input regarding your choice of ASP and the audit team make up. Sometimes it is better to spend a little more to get an audit team that better understands your operations.

4. Be prepared. Review the protocol, review your documents and have the materials you need to demonstrate compliance ready as well as having the people ready that will need to discuss the topic. Time spent looking for documents or waiting on personnel to become available is time and money (your money) wasted.

5. If you have an OMS, effectively map your OMS to the 17 Elements. While you and your personnel are used to discussing procedures and processes relative to your internal management system, the ASP personnel work in the structure of the 17 Elements of SEMS II. Audit time and resources should not be spent with the auditors and the operator's personnel trying to figure out where in the internal OMS a particular Element is addressed. There are many effective ways to map between your OMS and the 17 Elements, just do it ahead of the audit.

6. Provide audit "guides and translators". It is highly likely that the audit team provided by the ASP will include members whose background does not include offshore oil and gas operations. They will talk in terms of the 17 Elements, and will not understand all of the acronyms and slang utilized inside the industry. I have seen potential findings arise from an audit team member not understanding key points of what operations personnel were describing, or from operations personnel not understanding the auditor's question.

7. If you already know you have issues or missing items, admit it. Efforts to justify or minimize such items seldom do anything but take time away from productive discussions. Most auditors appreciate the fact that you have already identified an issue and are working to correct it. Admit it and move on.

The primary assessment of your SEMS program is the third party audit. Since you have to do it, and you have to pay for it; get the most for your money. Be prepared, be involved in the ASP selection, be transparent and learn from the results.

This wraps up the high level view of the what, why and how of the implementation of a BSEE SEMS program. For some of you this is enough, but for those of you who are on the front lines of

SEMS implementation it is time to take all the background information from Chapter 1 through Chapter 6 and head into Chapter 7, "The SEMS Elements", where the details of each element are discussed.

Suggested reading

Code of Federal Regulations, 30 CFR 250, Oil and Gas and Sulfur Operations in the Outer Continental Shelf, Subpart S, Safety and Environmental Management Systems (SEMS).

CHAPTER 7

The SEMS elements

Contents

Time to turn up the equipment and dig deeper. If you have looked at 30 CFR 250 Subpart S and API RP 75, you may be wondering how deep we can dig. Both of these references together are under 40 pages in total. That is exactly the issue we have to deal with. There is no handy checklist of required items to follow, because the references are describing a process not a set of rules. It is a management system, bringing with it all that implies; procedures, delegations, continuous improvement, etc. Understanding the goal of this management system requires understanding how and why SEMS was developed, what a management system is, and the relationship between SEMS, process safety and ISO 14001. Thus the earlier chapters spent on these topics. Remember, SEMS I and SEMS II both came in the aftermath of the Deepwater Horizon incident with the purpose to reduce the risks of future incidents. I don't use the term "prevent another incident" or the term "reduce the number of incidents" simply to avoid going down the rat hole regarding the never ending controversy and discussion that revolves around the question of incident elimination vs. reduction. If the term "prevent" is used it can elicit the debate regarding the potential for the "Black Swan" incident where the outcome was out of anyone's control. If the term "reduce" is used someone will comment that we should never settle for having any incidents. Behind both sides of this discussion is the core concept which is that of reducing risk.

There are three primary sources of reference material. I utilize regarding the SEMS elements:

- 30 CFR 250 Subpart S
- API RP 75
- COS-1-01

In the following sections, this book includes for each SEMS element the wording from the actual regulation which is 30 CFR 250 Subpart S. A copy of this can be obtained from the Federal Government Publishing Office. CFR 250 Subpart S references API RP 75 and multiple documents from the Center for Offshore Safety (COS). API RP 75 is a product of API and as such must be purchased from API. However, there is a link on the COS Website to access a read only copy for those not needing to have a personal copy. COS-1-01 is the SEMS Audit Protocol developed by the Center for Offshore Safety (COS) and is available on the COS website. Don't be intimidated by the nearly 40 page length or the approximately 400 questions contained in COS-1-01, the next sections will help you to understand what the core drivers within each element are from an operations perspective. Remember, however, when you are audited by an accredited ASP, COS-1-01 will be the protocol used. It is critical that you are familiar with the protocol. This will be discussed in more detail in Chapter 8, "The SEMS Audit", but COS-1-01 is not only useful in audit preparation but it also illuminates the components of each element in more detail than either CFR 250 Subpart S or API RP 75. In Chapter 6, "Getting SEMS Started" I discussed self assessments of your SEMS program. COS-1-01 should be an integral part of any self assessment.

The following sections on the individual elements are organized into three subsections. First will be the actual wording from 30 CFR 250 Subpart S (7-1-13 Edition). Remember, this is from the Federal Code of Regulations and can be revised so always check to be sure you have the latest version. Changes in 30 CFR 250 Subpart S tend to be relatively significant events, so if you are actively involved in the OCS oil and gas industry you will likely hear of proposed revisions before they take place.

Second will be a discussion regarding key concepts for operations personnel to understand within each element. This discussion purposely does not cover each item in the COS Audit protocol, not because they are not important but because my work in the SEMS area has shown a gap when it comes to operations personnel understanding what SEMS means to them. Remember, 30 CFR 250 Subpart S requires that each organization "appoint management representatives who are responsible for establishing, implementing and maintaining an effective SEMS program" as well as requiring periodic audits of your SEMS program by an accredited ASP. This means there are personnel within the organization who indeed need to understand every nuance of 30 CFR 250 Subpart S, API RP 75 and COS-01-1. And make no mistake; these personnel need to be integral to the operations organizations and provide the leadership and direction required by the regulations.

However, it has been my experience that in discussions with Asset Managers, Offshore Installation Managers (OIM), Persons in Charge (PIC), operations engineers, and other operations support personnel that SEMS can be envisioned as something that the HSSE or Regulatory Team handles. I have had an operations manager ask me to send him "all that SEMS stuff" so he could get familiar with SEMS. I explained that "all that SEMS stuff" is primarily contained in the processes and procedures that the organization wants followed in the course of daily operations. The only strictly SEMS document is probably the SEMS Program document whose primary purpose is to demonstrate how the organization is going to implement SEMS, and in many cases references the organizations internal processes and procedures. The following discussion sections are intended to mirror the discussions and training sessions I have had with various operations organizations to show them that SEMS is not separate from their daily operations.

Third will be a section labeled "what good looks like". I have taken my experience on both sides of the SEMS audit table and compiled a description of what I would view as key attributes of a thorough and rigorous implementation of that specific element. This compilation is based upon best practices I have observed as well as gaps I have observed. This is a bit of an element "dream team", in that it is comprised of the best of what I have observed, discussed or wished for as I have worked on the development, implementation and auditing of SEMS programs.

This requires a disclaimer in that the some of the organizations I have worked with have not always had internal processes or procedures to cover all the SEMS requirements. The first step for such organizations has to be to develop these processes and procedures. Only after that can implementation begin. Additionally, the following sections are not a guarantee that personnel will effectively answer every audit question, that your audit will have no adverse findings; nor are they a guarantee that an organization will not have any incidents.

What the following element discussions do intend to accomplish is to bring the understanding of SEMS out of the fog of regulatory requirements and into terms that help operations personnel understand how SEMS impacts them and the roles they play in implementation. Remember from Chapter 6, "Getting SEMS Started", the difference between SEMS and other regulatory requirements is that SEMS is based upon the implementation of a management process as opposed to being a set of rules with a pass/fail checklist.

7.1 Element one – general

Shutterstock #772014820

§CFR 250 250.1909 – What are management's general responsibilities for the SEMS program?

You, through your management, must require that the program elements discussed in API RP 75 (as incorporated by reference in § 250.198) and in this subpart are properly documented and are available at field and office locations, as appropriate for each program element. You, through your management, are responsible for the development, support, continued improvement, and overall success of your SEMS program. Specifically you, through your management, must:

(a) Establish goals and performance measures, demand accountability for implementation, and provide necessary resources for carrying out an effective SEMS program.

(b) Appoint management representatives who are responsible for establishing, implementing and maintaining an effective SEMS program.

(c) Designate specific management representatives who are responsible for reporting to management on the performance of the SEMS program.

(d) At intervals specified in the SEMS program and at least annually, review the SEMS program to determine if it continues to be suitable, adequate and effective (by addressing the possible need for changes to policy, objectives, and other elements of the program in light of program audit results, changing circumstances and the commitment to continual improvement) and document the observations, conclusions and recommendations of that review.

(e) Develop and endorse a written description of your safety and environmental policies and organizational structure that define responsibilities, authorities, and lines of communication required to implement the SEMS program.

(f) Utilize personnel with expertise in identifying safety hazards, environmental impacts, optimizing operations, developing safe work practices, developing training programs and investigating incidents.

(g) Ensure that facilities are designed, constructed, maintained, monitored, and operated in a manner compatible with applicable industry codes, consensus standards, and generally accepted practice as well as in compliance with all applicable governmental regulations.

(h) Ensure that management of safety hazards and environmental impacts is an integral part of the design, construction, maintenance, operation, and monitoring of each facility.

(i) Ensure that suitably trained and qualified personnel are employed to carry out all aspects of the SEMS program.

(j) Ensure that the SEMS program is maintained and kept up to date by means of periodic audits to ensure effective performance.

7.1.1 Understanding element one

Element 1 has two important characteristics that an effective management system needs to have to function. First, the SEMS program must be documented and that document available to operations staff, BSEE and audit teams. Two, SEMS needs to be an integral part of the Leadership and Management processes of the operations. Both of these warrant further discussion.

SEMS is a management system and as such there needs to be a guide to how an organization intends to implement the system. A management system needs a set of rules that all the participants are going to follow. For example, Management of Change (MOC) can be effectively implemented in a number of ways. There are sophisticated software packages that control who can initiate an MOC, who can approve an MOC, what information must be included in the MOC, etc. While these help control the process, never doubt the ability of individuals to develop short cuts and workarounds. There are also operators who use a handwritten MOC form and depend upon an MOC Coordinator to make sure the documented MOC process is followed. How an operator implements an MOC program is not nearly as important as what the operator determines to be the required steps, delegations and required information. In the off-shore world, every person on every shift of every hitch needs to do it the same way. Consistency in results is virtually impossible if there is not consistency in the process. And the start to consistency in process is when an organization writes down what that process is.

The document that sets out what these processes are that will make up an operator's SEMS is the SEMS Program document. Now there are no rules regarding what this particular document looks like or how it is organized. It just has to work for your organization. I have seen two types of SEMS Program documents, but that does not mean there are not others. The first type is becoming less common as operators' development of their SEMS programs becomes more established. This first type of Program document contains most of the details of how each element will be addressed by the operator. The actual processes are integral to the document. For example, the full details of the Hazard Analysis (HA) program are included in the SEMS Program document.

More commonly I see SEMS Program documents that direct the reader to the operator's detailed process documentation. Using MOC for an example, the MOC Section of the SEMS Program document may be as simple as referencing the process documents that detail how the operator will implement their MOC process. Since the process is already documented in detail via those internal documents, there certainly is no need to repeat it within the SEMS Program document. Just remember that all the referenced documents need to be available and understood by the operating staff. Referencing a document that the operating staff does not know exists or how to access it gets the same results as not having the document.

While this sounds easy enough, Element 1 gets its share of findings and noncompliances come audit time. You must have a SEMS Program document. You can't try to double duty or repurpose some other document the HSSE Department had regarding Control of Work (COW), process safety or something else. You need to produce a specific document. Some operators tie this document to the documentation for their larger scope Operations Management System (OMS), and that is fine. Just remember, the ASP audit Team thinks and works in a 17 element world.

The most interesting Element 1 failure I have witnessed, both as an auditor and as an Operations Manager, is where the Program document does not match the processes taking place in the field. To be clear, this does not mean the actual processes occurring in the field are wrong or dangerous. However, when there is a lack of consistency across shifts and hitches there is reduced control regarding operating risks.

At the risk of sounding cliché, if it does not get measured it does not improve. This is true for any management system. Simply put, your SEMS program needs to be treated the same as any other program, initiative, control of work, or financial objective with respect to management participation. SEMS should not be the subject of "special" meetings or managed by a separate "task group". It should be integral to the management processes in place. This means that the normal performance reviews and metrics will include SEMS items, and that improvement in operations' metrics is accompanied by improved SEMS implementation.

As an audit team member, I found an easy way to judge if this is the case is by simply asking to look at the daily, weekly, quarterly Management Team meeting agendas and topics. This is also a good internal self test. Are there metrics being reviewed by all levels of management that tie directly to improving the implementation and impact of your SEMS Program? COS-1-01 includes multiple questions regarding the dedication of resources to the SEMS effort, as well as questions regarding the support of management for SEMS. I would suggest that if SEMS is indeed integral to the way your operations are managed, such things will naturally take place. Metrics don't get presented and action items don't get assigned without resources dedicated and management support for the SEMS program. Conversely, if an operator reviews the SEMS program in a "special" quarterly or annual meeting it does not bode well for SEMS playing an important role in the everyday work processes. In fact, in those cases I like to ask for a record of who attends these special reviews, and many times find that the only regular participants are those directly tied to the SEMS program, and that the integration into the rest of the management team is limited.

So, Element 1, General, is really pretty simple. Develop your SEMS program, document it, implement it, and integrate it into the daily operations.

7.1.2 What good looks like

I look to two specific characteristics when judging Element 1. First is the SEMS program document, which is the foundation for everything to follow. It is relatively easy to review a good SEMS program document. Characteristics of such a document:

- The document is available and complete. While this sounds pretty basic, the document needs to be in final form and approved by the appropriate delegation of authority.
- The document follows the 17 BSEE SEMS elements or provides an easy to follow mapping from the SEMS program document elements to the 17 SEMS elements. It is a waste of resources and time if the audit team has to dig around in an operator's document library to attempt to find the documents that address a specific SEMS element.
- Documents referenced within the SEMS program document are available and easy to access. If the specific documents associated with the daily implementation of the SEMS program are difficult to access, it is hard to set an expectation of operations staff to follow them.
- The delegations for the responsibility for the implementation of the SEMS program and the management representative assigned to SEMS implementation are clear.
- The process for developing and reviewing SEMS implementation metrics is clear and is presented as a priority.

The second is the evidence that the SEMS program is seen as a priority and treated similar to other high priority objectives within the organization. When this is occurring, observations of the organization include:

- SEMS specific metrics are included in the normal performance and management reviews
- Participation in the SEMS metric reviews includes all key members of the leadership team, not just the ones assigned to SEMS

- Institutional knowledge of SEMS is evident, and appropriate for the individuals interviewed. This indicates that not only do the individuals have an understanding of the totality of SEMS, but more importantly they understand their specific role in it.

7.2 Element two — safety and environmental information

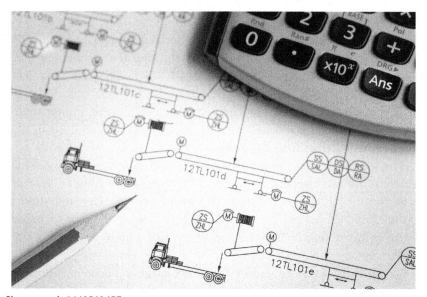

Shutterstock #440512657

§ 250.1910 — What safety and environmental information is required?

(a) You must require that SEMS program safety and environmental information be developed and maintained for any facility that is subject to the SEMS program.

(b) SEMS program safety and environmental information must include:

(1) Information that provides the basis for implementing all SEMS program elements, including the requirements of hazard analysis (§ 250.1911);

(2) process design information including, as appropriate, a simplified process flow diagram and acceptable upper and lower limits, where applicable, for items such as temperature, pressure, flow and composition; and

(3) mechanical design information including, as appropriate, piping and instrument diagrams; electrical area classifications; equipment arrangement drawings; design basis of the relief system; description of alarm, shutdown, and interlock systems; description of well control systems; and design basis for passive and active fire protection features and systems and emergency evacuation procedures.

7.2.1 Understanding element two

Here is a common story. The SEMS audit schedule has Element 2 scheduled for today. The element is "Safety and Environmental Information", and the HSSE Team is in the room, fully equipped with charts and tables showing injuries, near misses, spills, loss of primary containment data, stop the job forms, behavioral statistics, etc. The operator's personnel are comfortable and relaxed because they know this one is in the bag. They have enough HSSE data to satisfy any request. However, none of it has anything to do with what the ASP's Audit Team is looking for. Nobody in the room has actually read any of the SEMS reference material to understand what this element title refers to. Generally, the next step is the audit team lead stops the presentations and announces to the surprised gathering that this is not the information the audit team is looking for.

Element 2 is an easy one. But the title can be misleading. What this element requires is process and mechanical design data, operating specifications and limitations, etc. It is the information that is critical to developing and implementing a safety and environmental management system, not the outputs from your HSSE programs. This is a bit more understandable when one looks at the remaining SEMS elements. Without this information it is difficult to envision how many of the elements can be implemented, including (but not limited to):
- Hazard Analysis
- Management of Change
- Operating Procedures
- Training
- Mechanical Integrity
- Emergency Response and Control
- Incident Investigation

Once the requirements of Element 2 are understood and the information is gathered, the next challenge is easily accessing the myriad of information. I am always troubled when it takes personnel from operations,

IT, HSSE and engineering to locate the required information. It is essential that an operator have the information and that it is available across the organization such that it can be used on a daily basis. Human factors work tells us that there is a limit to just how many steps and how much time personnel are likely to spend obtaining information. I have found that in many cases once the operator understands what information is required to satisfy this element that the information trickles in as the audit progresses. The audit day after the discussion on Element 2 often has operators' personnel bringing in data books, design books, notebooks, thumb drives, boxes of files, etc. My experience is that more times than not the information exists, but it can take effort and resources to locate it. Don't underestimate the work this may require. As an operations manager I have found critical design information in the headquarters office, the shore base, on the facility, in the office basement, the acquisitions department file room, offsite storage (usually mislabeled), etc. Don't wait until two months before your next audit to look for it!

7.2.2 What good looks like

As I discussed above, it is a relatively simple exercise to understand what information needs to be a part of an organizations' safety and environmental information, and a rather tedious subsequent step of gathering it all. Once an organization has gathered all the information, the key to having information of any kind is the ability to access it. Once again, think of the Safety and Environmental Information as that information that is required for implementation of much of the rest of the SEMS program. This means that a significant number of people potentially require access to the information. Unlike a plant environment where all those needing the information are at the same site, the offshore world is much more complicated than that. The information likely needs to be available in the office, the offshore facility and the emergency command center just to name a few. A hard copy design data book tucked away on the shelf of the platform is not much good to a team in the onshore command center responding to an incident.

Obviously, in our world of digital data and electronic access to records this can be solved via technology. It is possible for there to be one set of information electronically available to multiple users and locations. Design of the system needs to take into account the support available at different locations. For example, IT support to help with data access is likely more available in the office than offshore. The logistics of the offshore world can also require an operator to have contingency plans for accessing the information. The prime example is slow or intermittent connectivity to offshore locations. I experienced this on a facility where the audit team asked to see a copy of a drawing while we were on the offshore facility. Someone on the operations team promptly pulled up the electronic data base and requested the drawing. After the final mouse click the operations staff suggested we have lunch because it would take that long for the drawing to download. Later they revealed the cabinet full of hard copy drawings which they actually worked from, and discussed the difficulty involved with keeping a hard copy library current and up to date. This illustrates why this is not a simple problem to solve. The common answer to poor connectivity to the data is to provide hard copy information. Maintaining hard copy data bases is a task that requires someone to physically remove old hard copies and replace them with revised hard copies.

Bottom line, a lot of people need access to this information and it needs to be easily accessed and safeguards in place to make sure only the current versions are in use.

7.3 Element three — hazard analysis facility level

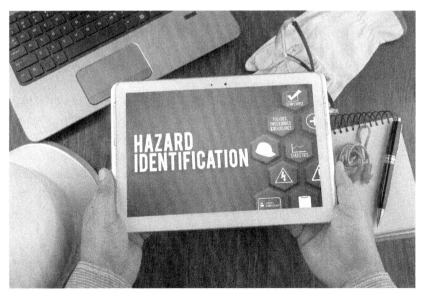

Shutterstock #520278556

The section of 30 CFR 250 Subpart S that addresses hazard analysis includes both the facility hazard analysis and task based hazard analysis. The section of 30 CFR 250 Subpart S included below is the section which addresses facility level hazard analysis. The section related to task level hazard analysis will be included in Section 7.4, "Task Level Hazard Analysis". This book will discuss facility level and task level hazard analysis as two separate topics.

§ 250.1911 – What hazards analysis criteria must my SEMS program meet?

You must ensure that a hazards analysis (facility level) and a JSA (operations/task level) are developed and implemented for all of your facilities and activities identified or discussed in your SEMS. You must document and maintain a current analysis for each operation covered by this section for the life of the operation at the facility. You must update the analysis when an internal audit is conducted to ensure that it is consistent with your facility's current operations.

(a) *Hazards analysis (facility level).*

The hazards analysis must be appropriate for the complexity of the operation and must identify, evaluate, and manage the hazards involved in the operation.

(1) The hazards analysis must address the following:

(i) Hazards of the operation;

(ii) Previous incidents related to the operation you are evaluating, including any incident in which you were issued an Incident of Noncompliance or a civil or criminal penalty;

(iii) Control technology applicable to the operation your hazards analysis is evaluating; and

(iv) A qualitative evaluation of the possible safety and health effects on employees, and potential impacts to the human and marine environments, which may result if the control technology fails.

(2) The hazards analysis must be performed by a person(s) with experience in the operations being evaluated. These individuals also need to be experienced in the hazards analysis methodologies being employed.

(3) You should assure that the recommendations in the hazards analysis are resolved and that the resolution is documented.

(4) A single hazards analysis can be performed to fulfill the requirements for simple and nearly identical facilities, such as well jackets and single well caissons. You can apply this single hazards analysis to simple and nearly identical facilities after you verify that any site-specific deviations are addressed in each of your SEMS program elements.

7.3.1 Understanding element three — facility level hazard analysis

I like to think of Element 3 as a "just get it done" element. It is that simple, but it is also that difficult. The self evaluation for this element is easy as well. Do you have a current, applicable and accurate facility level hazard analysis for all facilities that are operated by your organization? That is indeed a mouthful, and many times the answer is not a definitive yes or no. Just remember, the only answer an operator should feel comfortable with is "yes".

The first question most operators ask is what type of hazard analysis do I need? Does it have to be a HAZOP, HAZID, etc.? The answer, in what you will come to see is characteristic for SEMS is "it depends". There is a lot of wording in the SEMS associated documents that boils down to as the operator you are expected to assess your operations and develop a hazard analysis program that is appropriate for the complexity and risk of your operations. While this may be frustrating to those who like nice defined criteria, it is not reasonable to expect that a one size fits all solution exists for the diverse operations in the OCS. Since hazard analysis is the foundation for risk reduction, erring on the conservative side is preferable.

So what do you do? There is no substitute for combining risk analysis expertize with operations expertize and making an informed decision regarding the appropriate type of hazard analysis for your operations. My experience as an audit team member was that those operators with the most developed hazard analysis programs generally gravitated to the HAZOP process with a rigorous process for periodic reviews. Most risk analysis experts will also stress that the qualifications of the risk assessment facilitator is as critical to the results as is the selection of the method. Remembering back to Chapter 6, "Getting SEMS Started", the core purpose of SEMS is to reduce risk. Also remember that by itself a hazard analysis does nothing to reduce risk.

That is where the rest of Element 3 comes into play. The real progress is made when the results from the hazard analysis are understood and mitigation actions identified and implemented. Understanding the identified risks requires a method or process to prioritize risks such that those with the most significant impact may be addressed first. Most organizations use some type of risk matrix, be it a nine box, or twelve box or whatever. The key is utilization of a method to characterize risks that have varying degrees of probability and impact. Let me illustrate using some extreme examples. A direct meteor hit to a producing platform would indeed be catastrophic in impact, but the probability is extremely low. Not many platform risk mitigation plans include meteor avoidance. Conversely, it is highly probable that someone on the platform staff will get a paper cut this year. Yet you will not find any operations requiring gloves when handling written procedures. I agree, these seem a bit ludicrous, but think about where you rank things like supply boat impacts to the platform, pigging operations, exceeding of operating limits, etc. The results of a thorough hazard analysis can be very confusing and a bit intimidating without a process to categorize and prioritize. There is a temptation to solve this by saying that all hazards are equally important. This is a slippery slope as the logical follow up is requiring all hazards to be addressed immediately which has obvious resource implications. This is in no way any justification to allow high priority risk mitigation to be impacted by the resources available, but there is a need for effective prioritization and delegation utilizing personnel with the appropriate skill sets and experience.

At this point the organization has completed a hazard analysis and has prioritized the associated mitigation actions. The organization still has not reduced risk. To reduce risk changes must be made. With respect to any management system, the next step is quite standard. The mitigation action items must be assigned to owners who are responsible for their completion. Remembering Element 1, "General", the progress and completion of these mitigation action items must be tracked and reviewed on a timely basis, and prioritized appropriately for the risk involved.

This all looks pretty straightforward; complete a hazard analysis, identify risks, prioritize risks, develop mitigation action plans, complete action plans, However, Element 3 garners more than its fair share of audit findings and incident root cause credits. An effective hazard analysis program requires resources and discipline. The foundation is clear — you have to have a hazard analysis for all operations, regardless of the age of the operation, how it was acquired, or how the previous operator handled hazard analysis.

Hazard analysis is not a onetime deal where once you have done a HAZOP you can put it on the shelf and check that box "done". Any changes in the process or equipment have the potential to alter the hazard analysis, and should prompt a review of the hazard analysis. If a SEMS is working effectively, the Management of Change (MOC) process should trigger a review of the hazard analysis any time an MOC is executed. Additionally, SEMS requires that you have a documented hazard analysis review process that is appropriate for the risks associated with the operations. Typical review periods are five to ten years pending the risks associated with the operation. It is true that reviewing hazard analysis every five years can be a significant resource dedication, but an efficient MOC process which is prompting an organization to keep up as changes are made can go a long way to reducing the effort required at the periodic reviews.

Returning to the subject of risk prioritization I have seen hazard analysis action tracker listings that have hundreds of action items listed resulting from past hazard analyses. Invariably some of these action items have been on the list for years, which begs the question of how important are they? Does the long list of incomplete action items mean the associated risks still exist, or does it mean the list includes meteors and paper cuts? Eventually the organization gets accustomed to seeing these long lists and the importance of the completing the action items can be diluted.

Remembering that hazard analysis is the first step in risk reduction sets the tone for Element 3. Identify the risks, prioritize the risks, develop mitigation plans and then make changes to reduce risk. You have to make the changes to reduce risk!

7.3.2 What good looks like

An audit team member should be able to pick up a facility level risk assessment and without requiring additional documents or discussions understand the following:

- When was the hazard analysis completed and when is the next review scheduled?
- What type of hazard analysis was completed and why was that method chosen for the specific asset?
- What were the qualifications of the facilitator?
- Who were the participants and what were their qualifications?
- What risks and hazards were identified?
- How were the risks and hazards prioritized?
- What are the action plans associated with the mitigation of the risks and hazards?
- How will the action plans be assigned and tracked through completion (many times a separate action tracker will be used and the auditor will require access to that system).

As was mentioned above, there is no reduction of risk unless there has been action taken to mitigate the risks identified. Consequently, any individual with actions assigned to them should be aware they are the "owners" of these actions and be able to discuss the status of their completion. I have had the wonderful experience of setting down in an audit and seeing my name associated with an action item for the first time. And yes, the auditor asked me about it and my response was less than eloquent, followed up by a hallway "discussion" with the keeper of the action tracker.

During my time on the operator's side of the SEMS audit table, I also learned that the facilitator for the hazard analysis makes a significant difference. When specific facilitators submitted their final report, I could indeed answer all of the questions listed above. Many of us operations managers and engineers think we can facilitate a hazard assessment; until we see someone whose career focus is hazard assessment. There are industries external to oil and gas that are ahead in this area. Do yourself a favor and hire one of these people to see what a difference they make in the end product.

7.4 Element three — task level hazard analysis (JSA)

Shutterstock #551512852

§ 250.1911 — What hazards analysis criteria must my SEMS program meet?

This is section of § 250.1911, which addresses task level hazard analysis.

(b) *JSA*. You must ensure a JSA is prepared, conducted, and approved for OCS activities that are identified or discussed in your SEMS program. The JSA is a technique used to identify risks to personnel associated with their job activities. The JSAs are also used to determine the appropriate mitigation measures needed to reduce job risks to personnel. The JSA must include all personnel involved with the job activity.

(1) You must ensure that your JSA identifies, analyzes, and records:

(i) The steps involved in performing a specific job;

(ii) The existing or potential safety, health, and environmental hazards associated with each step; and

(iii) The recommended action(s) and/ or procedure(s) that will eliminate or reduce these hazards, the risk of a workplace injury or illness, or environmental impacts.

(2) The immediate supervisor of the crew performing the job onsite must conduct the JSA, sign the JSA, and ensure that all personnel participating in the job understand and sign the JSA.

(3) The individual you designate as being in charge of the facility must approve and sign all JSAs before personnel start the job.

(4) If a particular job is conducted on a recurring basis, and if the parameters of these recurring jobs do not change, then the person in charge of the job may decide that a JSA for each individual job is not required. The parameters you must consider in making this determination include, but are not limited to, changes in personnel, procedures, equipment, and environmental conditions associated with the job.

(c) All personnel, which includes contractors, must be trained in accordance with the requirements of § 250.1915. You must also verify that contractors are trained in accordance with § 250.1915 prior to performing a job.

7.4.1 Understanding element three — task level hazard analysis

First off, when SEMS talks about task level hazard analysis or JSA, it is essentially the same thing. A job safety analysis (JSA) is a commonly used and referenced task level hazard analysis tool. JSA is indeed included in Element 3, but it is important enough that it warrants its own discussion. I am sure most people reading this are thinking that JSAs are simple and wondering what the big deal is. The big deal is that as an audit team member I have observed a lot of findings for what should be an easy subject to comply with. There are two primary areas where operators fail to meet the SEMS JSA requirements.

First, know that there are some SEMS specific requirements that your JSA process must meet in addition to the basics of most JSA programs. So read the requirements included at the beginning of this section, understand them and implement them. Some of you will want to argue that some of these are not essential for an effective JSA program and once again — it does not matter. It is so much easier to comply than to protest and the requirements certainly do no harm. In my experience there are two frequently missed items, so be sure your JSA program:

- Requires the immediate supervisor of the crew performing the job onsite to:
  - Conduct the JSA
  - Sign the JSA
  - Ensure that all personnel participating in the job understand and sign the JSA
- Requires the Stop Work Authority procedures and expectations be documented as a statement on the JSA.

This does not mean that the immediate supervisor's lead technician can conduct the JSA and bring it to the supervisor. It does not mean that most of the crew has signed the JSA. It does not mean that you talked about Stop Work Authority instead of including it on the JSA form. It means just what it says.

Second, follow through on the JSA process for every JSA. Not just the big jobs or the "risky" jobs but all the jobs. Not completing the JSA process for all jobs can lead to a very slippery slope that you don't want to be near. I have reviewed a lot of very good JSA program documents, only to find that when I got to the facility they were not following the program. This is evident when the crew can't find a specific JSA, the JSA is perfectly clean and on the supervisor's desk and not at the job site, the completed JSA is a photocopy of yesterday's JSA, or when the JSA form is in English and the crew on site has limited English communications skills. These are just a few of the things I have observed. If the job involves a crew or shift change, the JSA needs to be appropriately reviewed and the new signatures added. Having the JSA signed by the previous crew does not satisfy any effective JSA program.

I have been surprised many times as an auditor when I have observed the simplest of all indications of the lack of utilization of JSAs, and yet the operations managers have not seen the same thing. The only answer is they are just not looking, and in most cases the suspicion is confirmed during the offshore portion of the audit. If a specific facility historically conducts 30 JSAs a month (see Section 7.15, Element 13 on record keeping) and that gradually becomes 15, I will ask if the activity level has changed significantly. If it hasn't it is a good bet that when you go offshore they are not following the operator's JSA procedures. And many times it is because the use of JSAs for "routine" work is not understood or clear. This is not to say that JSAs are a numbers game. The count will vary with activity level and that must be a consideration. Many operations managers, me included, have at one time set a "quota" for JSAs for a given facility. I can assure you the operations personnel met my quota every month, even if they had to do a JSA on the proper pouring of a cup of coffee.

7.4.2 What good looks like

This sounds so simple, but for some reason is so hard. Develop a JSA form that meets the organization's needs, and meets the SEMS requirements and then use only that form. And if you allow a contractor to use their JSA form while working for you, take time during the bridging process (as described in Section 7.8, Element 6, "Contractor Management") to assure it meets all the requirements as yours does, or make them use your form. Specifying which JSA form is to be used won't bring the work to a screeching halt; the importance is in the discussion and the process not the form. In the offshore world, there is a constant flow of contractors and individuals coming aboard, working and leaving. I am not sure, but I think my record as an auditor has been seeing at least four different JSA forms used on the same facility in the same month. Some met the SEMS requirements, and some did not. Once you are using a consistent form, it becomes easier to follow the SEMS requirements as listed above because these reminders can be built into the form. This does not guarantee a thorough JSA, which depends upon the discipline which the JSA process is implemented within the organization. Repeating what is included in the section above, follow through on the JSA process for every JSA.

A good JSA process is clear regarding when a JSA is required and when it is not. As an auditor I have seen many JSA program documents which are unclear, usually with respect to doing JSAs for "routine" work. This seems to lead to an adverse finding more times than not. When too much is left for interpretation, consistency in application is lost. If a JSA process does not require a JSA for routine work, individuals are left to determine what routine work is. This can lead to various decisions regarding what is routine work and when a JSA is and is not required. My advice is to eliminate the guesswork and require JSAs for everything.

Aside from the visual review of completed JSAs described above, a good JSA program can stand up to more in depth review. There is obviously a limit as to how many JSAs an audit team or operations manager can review. Incident investigation reports are another way to assess the discipline in the JSA process. As an audit team member I ask for a copy of random incident or near miss investigations. There should be a copy of the JSA included in the report, and the investigation team should have reviewed the JSA for thoroughness. Another approach is to review the JSAs associated with the work related to a specific MOC. It is surprising how many times the JSAs associated with incident investigations or MOCs are less than thorough. I can't really offer an explanation regarding why, but getting deeper into the JSA archives seems to reveal a lot.

If you think people are becoming complacent and just filling out the form without thinking about it, change the form, change your program, etc. But it is the operator's responsibility to assure that every time a JSA is conducted that it meets the SEMS requirements. This is not a guarantee of anything; the topic of JSAs can evoke discussions from both those who feel it is an integral part of any HSE management system and those who feel it is an ineffective routine. Aside from BSEE requiring that an operator utilize the JSA, it is still another tool in your management system toolbox so use it effectively.

7.5 Element four — management of change

Shutterstock #239359783

§ 250.1912 — **What criteria for management of change must my SEMS program meet?**

(a) You must develop and implement written management of change procedures for modifications associated with the following:

(1) Equipment,

(2) Operating procedures,

(3) Personnel changes (including contractors),

(4) Materials, and

(5) Operating conditions.

(b) Management of change procedures do not apply to situations involving replacement in kind (such as, replacement of one component by another component with the same performance capabilities).

(c) You must review all changes prior to their implementation.

(d) The following items must be included in your management of change procedures:

(1) The technical basis for the change;

(2) Impact of the change on safety, health, and the coastal and marine environments;

(3) Necessary time period to implement the change; and

(4) Management approval procedures for the change.

(e) Employees, including contractors whose job tasks will be affected by a change in the operation, must be informed of, and trained in, the change prior to startup of the process or affected part of the operation; and

(f) If a management of change results in a change in the operating procedures of your SEMS program, such changes must be documented and dated.

7.5.1 Understanding element four

Ok, get comfortable because this element is a critical one. In my opinion, Management of Change (MOC) is the heart of any operations management system. The long term health and performance of your management system depends upon an effective MOC process. Think about it for a minute. First you develop your SEMS Program and document the processes you intend to follow. Then you take the time to implement the processes and demonstrate to yourself and BSEE that you follow them. SEMS obligations completed, right? Wrong. The offshore world is a world of changes. Reservoirs decline, new wells are drilled, equipment breaks, technology advances, hurricanes occur, infrastructure ages, etc. And with every change comes the potential for your risk profile and hazard assessment to change. Change can impact operating limits, operating procedures, process flows and parameters, mechanical integrity requirements, just to name a few. How can we fully understand the full impact of changes, some of which may appear to be relatively benign at first look? The key is management of change. I have personally led serious incident investigations where the root cause was tied to poor execution of the MOC process.

Before going any further, I want to be clear that the success of an MOC process is not dependant on how sophisticated the electronic MOC tool being utilized is. Poor MOC performance is tied to lack of process discipline not the inability to afford the latest MOC software. This is because MOC is a process, not a form, not a formality. The process starts with documented, defined clarity of what does and does not require the MOC process to be initiated. When there is lack of clarity, human judgment and decision making are in control and even the best personnel are not infallible. In fact, I would propose that when in doubt; do an MOC as the process itself will clarify the need as you progress through it.

I also want to be clear that the purpose of this section is not intended to be a guide to developing a comprehensive MOC process. That would require a lengthy document in itself, and the best MOC processes I have observed have a high level of detail included in their MOC process documentation. When developing detailed management of change processes an organization needs to consider the types of changes, the significance of the changes and the appropriate level of review and approval. Rather, what I do intend to do here is to describe at a level of granularity similar to that of a SEMS audit team the evaluation of an MOC process. While having all of these items included in an MOC process does not in itself assure an effective MOC process, the lack of any of these items casts significant doubt regarding the MOC process. A poor MOC process has the potential to allow introduction of risk into the operations without fully understanding that risk. An effective MOC process has a few critical items which, not surprisingly, line up well with the SEMS Element 4 requirements in 30 CFR 250 Subpart S:

- A clear, technically complete description of the proposed change
- A completed and approved technical review of the proposed change
- A completed hazard analysis of the proposed change
- A comprehensive determination of the impacts of the change
- A determination of the personnel impacted by the change
- A comprehensive and completed communications or training plan for impacted personnel
- Review and approval of the change by the designated level of delegation
- A determination of those documents that require edits and changes and confirmation that the changes are made
- A process for pre start up review
- Clarity regarding who authorizes the startup

If you look at this list and your first reaction is how much work and discipline is involved, you are correct. This also makes the MOC process so vulnerable to short cuts and misuse. In an industry where downtime impacts can be measured in millions of dollars per day, it requires significant discipline to fully complete the process before taking the associated action. Again, this requires the organization to have clarity across the staff regarding what the MOC process is and what requires the MOC process to be initiated. The most basic MOC related failure is simply not knowing that a process to safely handle change even exists.

Aside from total lack of knowledge regarding the need to manage change, the failures are often some form of shortcutting the process. I have witnessed some very creative thought processes utilized to justify that a change did not require an MOC. If it takes several people several hours to justify that an MOC is not needed, odds are that an MOC is indeed warranted. Another waste of resources and creative thinking is developing this justification after the change is already been made. Remember, the goal here is to reduce risk. If you find a change that has been made without an MOC, and you think there is a chance that an MOC should have preceded the change, spend your resources reviewing the change and making any modifications your analysis shows to be required which includes shutting down operations. And don't trust that since the change was made years or months ago and nothing has happened that nothing will happen. That is not an acceptable form of hazard analysis!

Lack of closure is another common problem with MOC process implementation. By lack of closure I mean that not all the identified action items are completed but the associated change has been made. For any change that has been completed, there should exist an MOC document record which includes:

- A full description of the change to be made. One liners such as "install new separator" are not sufficient.
- Documentation of the technical review, which includes a sign off by the appropriate technical authority. Most MOC processes include the designation of required technical review and approval for MOCs.
- Documentation of the risk assessment results and the completion of any associated action items. The documentation should include the participants, the method used, the findings and recommendations and the delegation and completion of any associated action items.
- A formal indication of approval by the appropriate level of delegation within the organization. While this does not have to be a "wet ink"

signature, there must be some record of approval of the change. The fact that the change was routed to the appropriate approver does not constitute approval.

- Documents which have been edited or revised as a result of the change
- Documentation confirming the communication of the change. This is a common area of audit findings and process breakdown. Not every change needs to be communicated to all personnel and not all changes can be effectively communicated via the same method. Some changes can be communicated by an email, while some may require formal training. The MOC should include a communications plan and confirmation that is was completed. Accomplishing this can be a significant task considering multiple hitches, multiple shifts, vacations, illness, etc.
- Documentation of a pre start up review or justification why it was not necessary
- Documentation of the authorization to start up

Do not neglect document control with respect to MOCs. Regardless if you are using electronic or hardcopy documents it is important that when an MOC is reviewed and approved that all the items are included in the records. This is especially important in the event of a future incident investigation, but also holds true for an audit. For example, if a past MOC is requested as part of an incident investigation and the risk assessment cannot be found, it is essentially the same as not having done the risk assessment.

So, bottom line for Management of Change boils down to; describe the change, evaluate the change technically and with respect to the risk profile, complete identified pre change action items, communicate the changes to impacted personnel, revise impacted documents, and formally approve the change.

7.5.2 What good looks like

Because management of change is so important within a management system, and because it is so closely associated with identification of and mitigation of risk, it has been my experience that many operators do this well. This makes this section relatively easy to write as I have some good examples to use in this compilation. One characteristic stands out among all of those I consider effective MOC programs. Stealing a phrase from an old fast food advertising campaign, a good MOC program is the "full meal deal". When looking at an MOC from such a program, all the important information is included with the MOC. Regardless of whether it is a hardcopy file folder or an electronic MOC, if it is associated with the MOC process it is easily available. This is not just to make things easy for an auditor or an incident investigation team. Somewhere in the MOC process somebody is going to authorize the change to go forward. I cannot envision making the decision without having all the information in front of me. Another human factors consideration; if all the required information is not included with the MOC then that person authorizing the change must take the initiative to find the missing information. Ask yourself how often this will occur.

The reason for the change should be clear. This is more than simply saying what the change is; it explains why the change is necessary. According to 30 CFR 250 Subpart S, the management of change process must be followed when making changes to:

(1) Equipment,

(2) Operating procedures,

(3) Personnel changes (including contractors),

(4) Materials, and

(5) Operating conditions.

All of these can involve significant effort, resources and potentially bring new risk into the operation. Everybody involved should know exactly why the change is being made.

Because of the importance of understanding the impacts of the change the MOC should clearly show who reviewed the change, what their objective in the review was, what their conclusions were, and by what authority they were designated to the review the change. These are all pretty obvious to most people, with the exception of the last one. What does delegation of authority have to do with reviewing an MOC? The scope of changes that fall under the MOC process is very broad. Not every change requires the same level of expertize in the review process. It is important that an organization's MOC process includes guidance for this and the MOC should indicate how the reviewers were designated.

Changes can involve potential impacts to the process, impacts to the environment, and impacts to personnel. Such impacts need to be identified with rigor and thoroughness. Impacts can be positive, negative or neutral. A positive impact might be that a hazardous operation was automated to minimize human exposure. A neutral impact may be that the startup procedure for a new compressor is different than the old one.

Not any more hazardous or complicated, just different. A negative impact might be an increased risk of loss of containment. This begs the question of why would an organization make a change that has negative impacts.

The answer is they would not. However, without an effective hazard analysis of the change, negative impacts may not always be identified. The type of hazard/risk analysis performed for the change, who participated, the findings and the mitigating actions completed must be part of the MOC package. It is at this point that someone in the conference room blurts out something about spending more time on hazard analysis than producing oil and gas. I actually like that, because the response is so easy. Not every MOC requires the same degree of hazard analysis. Some may take minutes, some may take days or longer. Adding additional equipment and production to an existing facility probably requires significant time to complete a thorough hazard analysis. Refurbishing the galley needs a through hazard analysis but it is likely not going to require the same time to complete it.

Changes can impact personnel in many ways. For example, changes to operating procedures, changes to operating limits, changes to PPE, etc. Sometimes the change is significant, sometimes it is relatively minor. The key is to identify who is affected, how they should be informed or trained regarding the change and recording the evidence that the communications or training occurred. The recording part is not simply to satisfy the audit team. People are not perfect. People make mistakes even after training. During incident investigations training records are commonly requested. Enough said.

As we get to the final stages of the MOC process, some person is going to have to approve the change. There is no way around it. You can't automate or digitize the process of a designated person reviewing, understanding and evaluating the change. The MOC process needs to be clear how this designation is made and the MOC itself should indicate by what delegation of process this occurred. And in many organizations, the person who approves the change is not the same person who authorizes startup. This approve and authorize stuff can make your head hurt so using the addition of a new compressor to an offshore facility serves as a clarifying example. Assuming the MOC has everything completed and completed to the satisfaction of the person who has the delegation to approve making the change. The next step is the actual installation of the compressor. Once the actual installation is complete it is time for another human to make the decision to start the new compressor.

Enter the pre start up review (PSR). The best MOC programs I have seen have the PSR as an integral part of the MOC process. While not every change requires a PSR, the PSR or the justification for not doing one needs to be integral to the MOC. PSR is actually Section 7.11 Element 9, and the requirements for when a PSR is needed are discussed in that section. The installation of a new compressor would indeed require a pre start up review. Assuming the PSR is complete, somebody needs to make the decision to start it up. Who makes this decision is up to the organization. I have seen organizations where it is the offshore installation manager (OIM) and organizations where it is a member of the leadership team. Either way, the person on site and responsible for that facility needs to be involved in the decision as only they know the facility and the activities and conditions on any given day.

7.6 Element five – operating procedures

Shutterstock #100515940

§ **250.1913 – What criteria for operating procedures must my SEMS program meet?**
(a) You must develop and implement written operating procedures that provide instructions for conducting safe and environmentally sound activities involved in each operation addressed in your SEMS program. These

procedures must include the job title and reporting relationship of the person or persons responsible for each of the facility's operating areas and address the following:

(1) Initial start up

(2) Normal operations;

(3) All emergency operations (including but not limited to medical evacuations, and emergency shutdown operations);

(4) Normal shutdown;

(5) Startup following a turnaround, or after an emergency shutdown;

(6) Bypassing and flagging out-of service equipment;

(7) Safety and environmental consequences of deviating from your equipment operating limits and steps required to correct or avoid this deviation;

(8) Properties of, and hazards presented by, the chemicals used in the operations;

(9) Precautions you will take to prevent the exposure of chemicals used in your operations to personnel and the environment. The precautions must include control technology, personal protective equipment, and measures to be taken if physical contact or airborne exposure occurs;

(10) Raw materials used in your operations and the quality control procedures you used in purchasing these raw materials;

(11) Control of hazardous chemical inventory; and

(12) Impacts to the human and marine environment identified through your hazards analysis.

(b) Operating procedures must be accessible to all employees involved in the operations.

(c) Operating procedures must be reviewed at the conclusion of specified periods and as often as necessary to assure they reflect current and actual operating practices, including any changes made to your operations.

(d) You must develop and implement safe and environmentally sound work practices for identified hazards during operations and the degree of hazard presented.

(e) Review of and changes to the procedures must be documented and communicated to responsible personnel.

7.6.1 Understanding element five

Take a moment and forget SEMS and compliance and think about why operating procedures are important to a management system. Operating procedures are there to facilitate consistency among humans doing a task or a series of tasks. Operating organizations spend significant resources developing operating procedures that minimize and mitigate risk. Remember, we are dealing with humans when it comes to operating procedures, and the desired procedure may or may not be how any specific individual might approach the work if left on their own. Consequently, to reduce the overall risk profile every person on every shift of every hitch needs to follow the procedure that has been developed, reviewed and approved.

This is not meant to infer that operating procedures are a "one and done" exercise. Personnel will identify improvements to the procedures. Procedures must keep up with operating changes, process changes, etc. The primary prompt for operating procedure modifications should be the MOC process. Just remember that SEMS requires a review frequency and review process to be part of the operating procedures program. I have witnessed operators attempt the argument that an operating procedure is reviewed every time it is used, and therefore, a set period for review of operating procedures is not required. It is a far better use of resources to review your operating procedures at an appropriate interval. I have seen operators who have annual review periods and some with five year review periods. 30 CFR 250 Subpart S does not specify the review period only that it must be appropriate for the risks associated with the operating procedure. Regardless of what you decide and what you can support for each individual operating procedure the reviews can uncover changes in the facility or process that have not been captured in the current version of an operating procedure. My personal example is associated with a relatively large compression facility where produced gas was dehydrated and compressed to pipeline pressure. An internal auditor came to my office to review the results of his recent audit of the facility which was within my area of responsibility. The auditor was pleased to tell me that he asked multiple operators how to start up the facility after an emergency shutdown and they all stepped him through the same process. As I was starting to feel pretty good about this audit, the auditor told me that the problem was that the written operating procedure did not reflect what the operators had told him. My next step was to figure out how this had happened. What I discovered was that the steps as detailed by the operators were correct, and that they had edited a copy of the operating procedure to reflect changes in the process and sent it in to initiate the operating procedures review process over a year prior. Concurrently to their submitting the edited copy, there was a re-organization, roles changed and somewhere the edited copy got lost in the chaos. The operators grew tired of trying to find out where the new version was and just worked off their own process. So where is the risk introduced in this scenario? If one of the experienced operators was replaced by a new operator who went to the operating procedure library for the start up procedure, the procedure would not reflect current operations.

SEMS has some specific requirements when it comes to operating procedures, but none of that matters if the procedures are not used. Human factors work tells us this means the operating procedures must be documented, easily available to those implementing them, and in a format that is easy to follow. In general this means a check list type of format organized with the steps in the order they are to be completed. An operating procedure that is 30 pages of text will likely not find its way to the job site. At the completion of the work the associated procedures should be dirty, with the steps initialed and a supervisor's signature that the work is complete. Never trust a clean operating procedure. This sounds simple but it actually introduces some complications that have to be discussed. Operating procedures must be written for the intended implementers. Operating procedures written for competent operators are likely to be different than those written for newly hired personnel, both in detail and length. I have observed that most operators write operating procedures for competent operators. This then requires work process safeguards to prevent less competent individuals from implementing these operating procedures.

With respect to SEMS, the requirements specifically state that operating procedures must be in place to address:
- Initial startup
- Normal operations
- Emergency operations
- Normal shutdowns
- Startup following a turnaround or emergency shutdown
- Bypassing and flagging out of service equipment
- Safety and environmental consequences of deviating from equipment operating limits
- Properties of, and hazards presented by the chemicals used in the process
- Precautions that will be taken to prevent exposure of chemicals used in the operation, both to personnel and the environment

- Raw materials used and the quality control procedures used for purchasing the raw materials
- Control of hazardous chemical inventory
- Potential impacts to the human and marine environment as identified in your hazard analysis

This list causes a lot of angst and heartburn when the operator's definition of operating procedures and what they contain may not concur with this list. However, what is important is that somewhere within the operator's organization is a documented procedure or process that covers these items. This is best illustrated by examples. Many operators have a safe work practice which may be maintained by the HSSE team and implemented by operations regarding the handling and storage of hazardous chemicals on the facility. Just because the document is classified by the operator as a safe work practice and not an operating procedure is immaterial as long as the operations personnel follow it. Required personal protective equipment (PPE) associated with an operating procedure may not be listed within the operating procedure because there is a safe work practice requiring all personnel who enter into the unit being worked on to wear specific PPE every time they enter regardless of the purpose of the visit. Again, the important point is that there are documented requirements that are being followed. A tip for an effective audit is to know where in the organization all these "procedures" reside and be ready to access them.

The topic of operating limits and consequences of deviation generates a great deal of discussion. As written, SEMS requires this to be part of the operating procedures. As control technology progresses, many operators have this information available via the facility control system. The key is to be able to demonstrate that the operations personnel understand the significance of this information and that it is readily available to them and maintained. A word of caution is warranted. If the primary source of this information is via the control system interface, it is good to have some kind of backup provisions in the case where the online system is not operational.

7.6.2 What good looks like

The true test of any operating procedure program is the evidence that the operating procedures are being used. To be used the operating procedures must be easily accessed by the operating personnel. If it takes help from the IT folks to get into the electronic operating procedures library it is a good bet the operators are not spending much time reviewing them before they undertake the work. An indication that this is going on is the operators having an outdated hard copy of the operating procedure that they use because it is too difficult to access the current one.

The most used operating procedures are in a checklist or task based format. If they are written for competent operators the detail needs to be sufficient to assure understanding yet concise enough that operating personnel can be expected to use them consistently. I have seen some very good operating procedures that highlight cautions and hazards associated with specific steps. The checklist needs to be easily reproduced so it can be taken to the job site and the steps followed in the correct sequence. When it is appropriate, there should be a space on the operating procedure where the operator can initial the completion of each step. Before you get concerned over this statement, I agree that there are different levels of risk associated with specific operating procedures. It may be appropriate to eliminate the need to initial each step in an operating procedure that has minimal risk associated. If you do implement such a system, be sure to clearly define the risk tolerance or categorization for which initialing the steps is and is not required.

A good operating procedure is also an important record. After the operating procedure has been implemented, the copy with the appropriate initialed steps should be signed by the appropriate supervisor and the completed operating procedure filed for reference. I am sure somebody is reading this and thinking 30 CFR 250 Subpart S does not require operating procedures to be maintained. You would be correct; there is no SEMS requirement to save executed operating procedures. I recommend this step for two reasons. First, this provides a method by which the supervisor can assure that the operating procedures are being used. Second, in the event of an incident the completed operating procedure can be very helpful in determination of the root cause.

Bottom line for operating procedures is if they are not being used, they might as well not exist. Good operating procedures are being used consistently as part of how work gets completed on a facility.

7.7 Element six — safe work practices

Shutterstock #702935881

§ 250.1914 — What criteria must be documented in my SEMS program for safe work practices and contractor selection?

Your SEMS program must establish and implement safe work practices designed to minimize the risks associated with operations, maintenance, modification activities, and the handling of materials and substances that could affect safety or the environment. Your SEMS program must also document contractor selection criteria. When selecting a contractor, you must obtain and evaluate information regarding the contractor's safety record and environmental performance. You must ensure that contractors have their own written safe work practices. Contractors may adopt appropriate sections of your SEMS program. You and your contractor must document an agreement on appropriate contractor safety and environmental policies and practices before the contractor begins work at your facilities.

(a) A contractor is anyone performing work for you. However, these requirements do not apply to contractors providing domestic services to you or other contractors. Domestic services include janitorial work, food and beverage service, laundry service, housekeeping, and similar activities.

(b) You must document that your contracted employees are knowledgeable and experienced in the work practices necessary to perform their job in a safe and environmentally sound manner.

Documentation of each contracted employee's expertise to perform his/her job and a copy of the contractor's safety policies and procedures must be made available to the operator and BSEE upon request.

(c) Your SEMS program must include procedures and verification for selecting a contractor as follows:

(1) Your SEMS program must have procedures that verify that contractors are conducting their activities in accordance with your SEMS program.

(2) You are responsible for making certain that contractors have the skills and knowledge to perform their assigned duties and are conducting these activities in accordance with the requirements in your SEMS program.

(3) You must make the results of your verification for selecting contractors available to BSEE upon request.

(d) Your SEMS program must include procedures and verification that contractor personnel understand and can perform their assigned duties for activities for activities such as, but not limited to:

(1) Installation, maintenance, or repair of equipment;

(2) Construction, startup, and operation of your facilities;

(3) Turnaround operations;

(4) Major renovation; or

(5) Specialty work.

(e) You must:

(1) Perform periodic evaluations of the performance of contract employees that verifies they are fulfilling their obligations, and

(2) Maintain a contractor employee injury and illness log for 2 years related to the contractor's work in the operation area, and include this information on Form BSEE−0131.

(f) You must inform your contractors of any known hazards at the facility they are working on including, but not limited to fires, explosions, slips, trips, falls, other injuries, and hazards associated with lifting operations.

(g) You must develop and implement safe work practices to control the presence, entrance, and exit of contract employees in operation areas.

7.7.1 Understanding element six — safe work practices

Before jumping into what a Safe Work Practice is or is not, there is a unique aspect of to this element that needs to be clarified. As you read through the regulations and documents, you will notice that a large portion of this element deals with contractor management. Why this is combined with Safe Work Practices is not important, just be cognizant that it is in here. Contractor management is important enough that it is covered in Section 7.8 as a separate topic.

Back to the first part of Safe Work Practices (SWP), what are safe work practices? That is a very good question, one that has been discussed across audit tables at great length usually to no constructive end. My answer is "who cares"? You can call it a SWP, a procedure, a process, a work rule, etc. What is important is that at a minimum there is a documented method which is followed that details effective and appropriate methods for:

- Opening pressurized or energized equipment or piping;
- Lockout and tagout of electrical and mechanical energy sources;
- Hot work and other work involving ignition sources;
- Confined space entry;
- Crane operations;

While these are the specific SWPs called out in 30 CFR 250 Subpart S via the inclusion of RP 75, it is likely that your organization has other SWPs as well. These can include everything from PPE to barricading requirements.

It does not matter if these are covered in five documents or fifty documents. You must have documented, effective methods and be able to demonstrate that they are indeed being followed. Once again, let's get back to basics. The goal is risk identification and mitigation, not having SWPs in a book on a shelf. Think about these topics for a minute. You would be hard pressed to find a hazardous task that does not require one or more of these SWPs. You must get this right! This brings me to what I see as a common finding when reviewing SWPs; out of date documents. It is imperative that you review SWPs at regular intervals, as well as when impacted by changes (which should involve an appropriate MOC, the "heart" of a management system). The operations manager side of my brain does not like the idea of putting my trust in an old SWP document.

It is probably no surprise that along with these SWPs there must be an effective permit to work system (30 CFR 250 Subpart S via the inclusion of RP 75). Per RP 75, a permit to work must be issued before work involving the following areas is initiated:

- Opening pressurized or energized equipment or piping;
- Lockout and tagout of electrical and mechanical energy sources;
- Hot work and other work involving ignition sources;
- Confined space entry;

Offshore operations are complex and contained within a limited footprint. There is no room for error. It is not enough to follow the SWPs, your permit system must effectively bring together the SWPs with the appropriate personnel to provide the assurance that the work being done has been fully reviewed, all mitigations implemented and the work plan communicated effectively.

It is not uncommon for multiple work sites to exist at the same time on a given facility, a situation referred to as simultaneous operations (SIMOPS). It is critical that an operator have a SWP that addresses how this will be reviewed and permitted on any given facility. The SIMOPS SWP may be simple for smaller operations and very detailed for larger operations. I have seen small, unmanned platforms where SIMOPS are discouraged but in the event they are required operations, engineering and operations will develop a specific plan for that situation and present it to management for approval prior to starting work. In larger facilities where it is a more common occurrence there is detailed SIMOPS process, many times including a SIMOPS review meeting as a part of the facility's daily cadence of meetings. Of note is the fact that while 30 CFR 250 Subpart S does not specifically mention SIMOPS, the audit protocol developed by COS does (COS-0-01 question 690g).

Material handling is also contained within this element, and it can be the subject of multiple issues and in my opinion requires the use of subject matter experts (SMEs) to assure a comprehensive set of SWPs. This topic is complex, changing and requires support from personnel whose focus is materials handling and storage. 30 CFR 250 Subpart S includes wording regarding toxic or hazardous materials and there are also references to work practices meeting the most current provisions of federal, state or local regulations or flag state practices. This incorporates a lot of regulations and criteria and requires effective communication between those implementing the SEMS and those SMEs responsible for understanding and assuring compliance with all materials handling, storage and disposal regulations. In one sense, this may sound easy for those developing the SEMS. Incorporate all the processes and procedures developed by those SMEs responsible for the proper handling of materials and the compliance with the associated regulations. However, we all have seen simple become complicated when it involves numerous teams and responsibilities. The SEMS pitfall comes in assuring that the procedures and requirements developed by the materials handling SMEs are incorporated into the SEMS program and fully implemented into the field operations. For example, if an operator has a detailed SWP for the storage of hazardous materials, but the operations personnel have not been trained in what materials are classified as hazardous and which are not, there is potential for hazardous materials being found improperly stored or disposed of, creating an unnecessary risk. This is not limited to large quantities of materials used in bulk on the facility. Auditors and inspectors will look in closets, on shelves, etc. for improperly stored materials.

As an auditor the first clue that there is a potential for an issue is when the interview with the regulatory personnel demonstrates a through library of material storage and handling procedures and requirements, but none of them are referenced in the SEMS program and the operations staff does not know how to access these procedures.

7.7.2 What good looks like

There are many parallels between what a good SWP program looks like and what a good operating procedures program looks like. It all comes down to the operations staff following the most current safe work practices consistently and thoroughly. This means the current SWPs are easily accessible to those who need to follow them. Similar to operating procedures, if it is difficult to access the current SWPs, hard copies of outdated SWPs will begin showing up in operator's desks and stuck to the walls in the control room.

However the SWPs are stored, digital or hard copy, the library needs to be well organized, current and focused on those SWPs that are a priority to be followed. I have seen SWP libraries grow so large that the listing of SWPs is multiple pages. Sometimes there are SWPs that were written for something that was done once or is done every 10 years. Incident investigations can lead to the generation of SWPs to prevent recurrence of an issue when a change in an operating procedure or a design requirement is more appropriate. It is very easy to see the solution to many things as being the development of another SWP. I am not here to tell you what should and should not be in your SWP library. I am warning that if the list of SWPs grows too large the truly critical SWPs can get buried in the library. As you add SWPs to the library, ask yourself if the best solution to the problem you are facing is an SWP or is it more appropriate to address it via another approach.

As always, the rubber meets the road when it comes to evidence that the SWPs and permit process are being followed. Using lock out tag out (LOTO) as an example, the records need to be completed and maintained with discipline. If an auditor sees LOTO tags and locks on a piece of equipment it should be easy for the auditor to review the onsite records that show when and why the equipment was locked out, who authorized it, what risk mitigations were put in place, and who must authorize the removal of the LOTO locks. Conversely, if an auditor reviews an open LOTO permit they should not observe this equipment back in service on the facility visit. This sounds easy enough but when there is activity that requires LOTO, there may be pressure to get equipment back in service and it is easy to neglect the process. Never assume a contract crew understands the LOTO process.

The same is true with materials handling. The facility chemical inventory sheet should list every chemical on the facility, regardless of whether it is being used currently or not. I have seen chemicals leftover from a project, or that are no longer being used in the process still on the facility but removed from the inventory list. Additionally, the Safety Data Sheets (SDS) for each chemical on the facility should be easily available on the facility. Then there is the issue of the organization of the chemicals and materials on the facility. Along with proper training of personnel, clearly labeling what can and cannot be stored in specific areas is important. Despite the best efforts at training all employees and contractors, a little bit of signage may prevent someone from disposing of hazardous waste in a non hazardous waste container and the subsequent effort required to rectify this.

The SEMS Coordinator and the Regulatory Team need to work closely to make sure that the effort and resources put into understanding and complying with the regulatory requirements is translated into practice on the facility.

7.8 Element six — contractor management

Shutterstock #784578568

7.8.1 Understanding element six — contractor management

As noted in Section 7.7, contractor management is included in Element 6, Safe Work Practices. As an audit team member, I find this to be one of the most likely topics to result in a finding. As an operations manager, I am going to focus on the part of 30 CFR 250 Subpart S that is specific to managing contractors who have already been selected to do work for the operator. Do not neglect the part of 30 CFR 250 Subpart S that discusses contractor selection. My experience has been that with the tools and information available today most operators are able to develop a contractor selection process that includes the criteria in 30 CFR Subpart S. I will focus on managing those contractors once they have been selected and show up at the job site.

There is one big "nugget" to remember for this; the operator is accountable for the actions and qualifications of all personnel on the facility. The goal is risk reduction and it does not matter who individuals work for. Consequently, as an operator you are responsible for selecting contractor personnel who are qualified for the tasks they will be doing and assuring that they work in a safe and environmentally sound manner.

As SEMS is currently configured, the compliance with SEMS by contractors is the responsibility of the operator. To accomplish this, an operator needs to dedicate significant, qualified resources to contractor management. This goes far beyond using HSSE information to select a contractor and beyond the basic bridging document. Many contractors are developing their own SEMS programs, which is a significant benefit to the operator. However, the operator must still assure that the contractor's SEMS meets or exceeds the operator's requirements and meets BSEE requirements. It is also the operator's responsibility to assure that the contractor personnel truly implement the SEMS program regardless of whose plan it is. If this sounds like a lot of work for the operator; it is!

There are many ways for an operator to meet the requirements, and it may vary from contractor to contractor depending on size, capabilities, duration of contract, etc. For example, an operator may supply the personnel to be on site to assure the SEMS is followed, or may require the contractor to supply the personnel to do this and subsequently report to the operator's facility management. Just remember, the operator is responsible.

If a contractor does not have a SEMS program of their own, the operator needs to train the contractor's employees regarding the operator's SEMS program. This can seem daunting but remember that contractor personnel may only be impacted by a few of the SEMS elements so there is no need to train them in all 17 elements and make them experts in the operator's full SEMS program. For example; contractor personnel may be required to work within the operator's permit to work system and use JSAs, but will not be executing any operating procedures. While this seems straightforward, this is an area full of opportunity for problems. Contract personnel are often moved from one task to another as workloads are balanced and new work is identified. Some jobs involve multiple shifts and multiple hitches. Contract personnel get sick and take vacation. Someone has to track each individual, what they are working on and their individual qualifications to provide assurance that contract personnel are always assigned appropriately.

With more contractors developing their own SEMS programs, the concept of the bridging document is becoming more important. While a contractor having a SEMS program generally makes SEMS implementation by the contractor more efficient, it does not remove the accountability for the actions and qualifications of contractor personnel from the operator. This is where the bridging document comes into play.

A bridging document is intended to "bridge" between the contractor's SEMS program and the operator's SEMS program. To do this effectively, the two SEMS programs must be compared on an individual element basis. Where the programs are different, the bridging document needs to specify which will be used. A one page document stating that in the case of disagreement the most conservative or strictest practice shall be used is not an effective method to accomplish this. The "rules" need to be clear prior to the work beginning. Once the bridging document is complete, both operator and contractor personnel need to understand the requirements contained within it. For example, if the JSA forms are not the same, it needs to be made clear in the bridging document which form will be used on the job.

Once the "rules" are fully understood by all parties and work begins, someone needs to assure that the contractor is indeed following the rules as agreed upon. You guessed it — this is the responsibility of the operator. As with any procedure, process or program, simply having it in writing does not assure implementation. Having a thorough bridging document does not remove any of the accountability from the operator. Just because you told the contractor how you expect them to work, it is still the operator's responsibility to assure it is happening. A common audit finding is the lack of any process to confirm that contractor personnel are working as the operator has requested. Bottom line is that as the operator you do not want to be in an audit or incident investigation and discover that a contractor you selected was not working in a safe and environmentally sound manner.

7.8.2 What good looks like

For a lot of offshore facilities, it is a rare day when there are no contractors on board in some capacity. It then falls to the operations organization to manage these contractors and to assure the compliance with SEMS. When done well, the process to accomplish this is clear and documentation exists to demonstrate the effectiveness.

If the contractor has no SEMS program, then the contractor will follow all applicable sections of the operators SEMS plan. Recognize that there may be some elements that impact the contractor more than others and are the most important from a risk identification and mitigation focus. A maintenance contractor may have no need to understand or use any operating procedures, but will need to be very familiar with the JSA and Safe Work

Practices elements. In such cases there should be documentation of the training of the contractor personnel regarding those parts of the operator's SEMS program which the contractor is expected to comply with.

As previously noted, many contractors are developing their own SEMS program making the most common method of coordination of and compliance with SEMS between operators and contractors is the use of a bridging agreement. A good bridging agreement should be specific to the contractor and the work being done. Building on the previous section, the bridging document should be clear which elements of which organization's SEMS program are going to be used, and this requires an individual element review. It is important to look at each element in the contractor and operator SEMS program and be clear with respect to each element whose program is to be used. I see a lot of bridging agreements that include the general statement that if the contractor and operator SEMS programs do not agree, the more conservative will be utilized. The problem with this is it requires someone in the middle of getting the work done to make a decision regarding which program is most conservative. Whose definition of most conservative is used? This can vary depending on the individual evaluating it. It is much better to have it clearly set out in the guidelines before the work starts. This needs to be clearly documented as part of the bridging agreement and the training of the operator's and contractor's personnel associated with the work covered by the bridging agreement should be documented. The best bridging agreement I have seen had a section for each of the 17 elements, and detailed which organizations SEMS program would be followed for that element. The record of the training was attached to the bridging document as well as a signature page for the individuals involved in the training.

Next comes the hard part; making sure the contractor is actually following the bridging document. While this may sound simple, it is the source of a significant number of adverse findings. As an auditor or an operations manager, I ask contractors some basic questions such as:

- Whose JSA form are you using?
- Whose permit process are you using?
- Whose LOTO and hot work practices are you following?
- Who is the Ultimate Work Authority on this facility?
- What is the Stop the Work process on this facility?

Remember, many of the contract personnel work for a variety of operators and not everyone has the same SEMS implementation requirements. If any of these questions result in anything but a clear answer, it is usually a clue that the bridging document is either not detailed or has not been communicated to the field staff. Conversely, I have seen contractors who have the bridging agreement, know exactly what SEMS processes are to be followed and likely have an opinion regarding who has the best SEMS program in the GOM.

But being good at this does not fall totally on the contractor. Remember, the operator is ultimately responsible for the contractor's compliance with SEMS, regardless of whose SEMS program is being followed. This means the operator has a plan and resources dedicated to assuring this is happening. This often takes the form of a periodic audit of the contractor, a regular meeting to review SEMS implementation, or in cases where the level activity warrants, an operators representative on site with the contractor assuring compliance. The important part is not how this is accomplished but that it is being accomplished and that there is documentation to demonstrate compliance.

While it is tempting to complete the bridging document, stress your expectations of SEMS compliance to the contractor and then move on to something else, that is not enough. The best operators continue through the work in some way to assure that the SEMS requirements are indeed being met.

7.9 Element seven — training

Shutterstock #249640810

§ 250.1915 — What training criteria must be in my SEMS program?

Your SEMS program must establish and implement a training program so that all personnel are trained in accordance with their duties and responsibilities to work safely and are aware of potential environmental impacts. Training must address such areas as operating procedures (§ 250.1913), safe work practices (§ 250.1914), emergency response and control measures (§ 250.1918), SWA (§250.1930), UWA (§ 250.1931), EPP (§ 250.1932), reporting unsafe working conditions (§ 250.1933), and how to recognize and identify hazards and how to construct and implement JSAs (§ 250.1911). You must document your instructors' qualifications. Your SEMS program must address:

(a) Initial training for the basic wellbeing of personnel and protection of the environment, and ensure that persons assigned to operate and maintain the facility possess the required knowledge and skills to carry out their duties and responsibilities, including startup and shutdown.

(b) Periodic training to maintain understanding of, and adherence to, the current operating procedures, using periodic drills, to verify adequate retention of the required knowledge and skills.

(c) Communication requirements to ensure that personnel will be informed of and trained as outlined in this section whenever a change is made in any of the areas in your SEMS program that impacts their ability to properly understand and perform their duties and responsibilities. Training and/or notice of the change must be given before personnel are expected to operate the facility.

(d) How you will verify that the contractors are trained in the work practices necessary to understand and perform their jobs in a safe and environmentally sound manner in accordance with all provisions of this section.

7.9.1 Understanding element seven

Training could be either a short or a very long section. In the shortest version, the operator is responsible to make sure that all personnel, including contractors, are trained appropriately for their job responsibilities to work in a safe and environmentally sound manner. The intent of this concept is hard to argue with, but how this is accomplished and demonstrated is complicated in so many ways. This begins with the basic training all individuals need to simply travel offshore and be on a facility and proceeds into significant detail regarding the processes and procedures that are required to be followed and implemented to maintain and operate the facility. And of course, it needs to be documented. This section needs to be treated much like the old joke about how to eat an elephant — one bite at a time. There are several critical "bites" to understand.

To begin with, an operator needs to determine who needs to know what. Job descriptions and responsibilities can form the basis for what does an individual in a specific role needs to be qualified to do. Every person on the facility cannot possibly be an expert on everything required to operate the facility. It is not necessary nor is it practical. Obviously, when personnel are in training they are not available to operate the facility, which results in the much discussed resource cost of training. While selective training can reduce the amount of training hours personnel are required to complete, it does put additional accountability on the facility leadership. Leadership must provide assurance that personnel are not assigned to work in an area or on a task for which they are not qualified. This is harder than it might sound when you consider that some platforms will have several hundred people aboard at any one time and multiple projects ongoing with multiple contractors and staff. There is plenty of opportunity to move personnel between job sites without confirming that their qualifications meet the needs of the work.

After you have determined who needs to know what, then comes the question of how are you going to administer the training? This is not as simple as contracting with a training company and begin sending personnel to an offshore worker training program. While there are plenty of training companies available who can offer significant help in developing and delivering your training program, some of the training will be specific to the facility where the individuals work. This means there is a required step of determining the curricula for the specific roles. This in itself has taken significant resource dedication by those operators who have completed this step. Just to make this even more complex, the curricula can be organized in multiple ways. For example, one operator may organize by individual skill sets, such as how to replace the plate in an orifice meter. Another may organize by all the procedures measurement technicians need to be proficient at all tasks required to maintain orifice meters, one step of which is changing an orifice plate. There is no right or wrong way, but the operator needs to provide assurance that the team working on the meter are trained in all aspects of the work they have been assigned.

The training of contractor personnel warrants discussion as well. There are significant numbers of contract personnel working across the Gulf of Mexico and in roles ranging all the way to facility leadership. As an operator by your specifying the requirements you expect in the personnel provided does not remove you from having accountability that the personnel supplied meet these requirements. Before selecting a supplier for personnel, the operator should review the contractor's training program and documentation to assure that it meets internal and external requirements for type of training and the documentation of that training. But you are not done yet. Once contract personnel arrive on the facility, the operator needs to confirm that the individual personnel supplied have the training and qualifications as specified. It has surprised me as an operations manager when despite all of the planning and efforts of the operator and the contractor how many times unqualified personnel still end up arriving offshore or at the travel facility waiting to go offshore.

The next bite of the elephant is documentation of training. Regardless if you organize your training by role, procedures, tasks, etc. the operator needs to be able to demonstrate (there is that word again) that personnel, both company and contractor, have indeed been trained appropriately for what they are being asked to do. There was a time if the OIM observed an individual executing an operating procedure and the individual completed it correctly, the OIM gave their consent and the person was deemed qualified. Those days are long gone. There needs to be documented records for individuals containing their training history and individuals have been requested to demonstrate their qualifications as part of audits, inspections, etc. Before you get riled up and head down the road that so many have traveled declaring that the responsibility for determining qualifications rests with the operator and how it is accomplished is nobody's business consider this. During any incident investigation the qualifications of the personnel involved will be discussed. The training records can go a long way in ruling out root causes associated with personnel qualifications. Conversely, the investigation can uncover a gap in the training program that needs to be addressed. Without training records this is not possible.

Some of you will be cursing at this section wondering when it will address training vs. competency. I have left it until the end because in my opinion, which you may disagree with, "it just doesn't matter" (Bill Murray quote from the movie Meatballs). The bottom line is that the operator is accountable for all personnel on the facility being capable of doing any work they are asked to do in a safe and environmentally sound manner. You can call this competent, qualified, trained, etc. The semantics don't matter, but the program in place to assure that personnel are not assigned tasks that they are not capable of completing in that safe and environmentally sound manner does matter and matters very much.

7.9.2 What good looks like

While the understanding and implementation of an effective training program is like eating an elephant, describing what good looks like is relatively simple. In fact, it can be summarized in two words; documentation and discipline. This does not mean that accomplishing it is easy, but it is easy to describe.

Documentation is a recurring theme in what good SEMS programs look like, and training is no different. At anytime, on any facility documentation should exist for every person on that facility with respect to their training records. This means for every person, specific to that person and not for the role the person is in. As an auditor I should be able to see a name on a JSA, work permit or even a name on a hard hat and ask to see that specific person's training records. And the records need to be current. This is a source of many discussions across an audit table or in a project update meeting. Somebody who is critical to the work being done has not completed their required training. There is an easy solution to this; the individual cannot work in that role until the training is complete. A good operator catches this before the person is on the job site while they have the opportunity to find a qualified replacement.

I will throw a bit of real world common sense into this, because I am not so naïve to think that exceptions do not occur. There have been many an OIM or PIC upset because they were informed that somebody currently on the facility has a gap in their training. It is surprising how many times someone's training renewal date is mid hitch. I attribute this to people glancing at the training records and looking primarily at the year and month. What I am going to stress is that exceptions are the subject of leadership review, risk assessment and formally approved via the MOC process. I am not in any way endorsing allowing individuals to work in areas where they are not trained and competent. That is not acceptable, and if their training is not current for the work they are there specifically to do, they cannot work.

However, not all training is associated with equivalent levels of risk, and the best managed training programs fall short sometimes due to vacations, personal time off, etc. For example, a person whose training is not current for LOTO or hot work should never be allowed to work in a role where this is required. Doing LOTO or hot work incorrectly can introduce significant risk. On the other hand, if a production operator's first aid refresher (which they have taken on schedule for 20 years) due date turns out to be in the middle of their current shift, I would not necessarily swap them out mid shift. If the facility had sufficient first aid coverage via other individuals on board, I would consider completing an MOC to allow this person to remain onboard and requiring the refresher be completed prior to the next hitch. The alternative of removing the individual from their shift as an operator would increase the work hours of the remaining operators which could introduce more risk than allowing the individual to remain in their role This should be documented in the MOC and the MOC should have to be approved at a relatively high level in the company and should document any limitations placed on the individual.

Let's shift the discussion to discipline. That same individual identified from the JSA, work permit or hard hat and whose training records have been accessed should only be found working in those roles and assignments for which they meet the training requirements set out by the operator. While this also sounds easy it is a source of many adverse findings. If someone with a specific set of skills and qualifications is needed on the facility and no one on the facility meets the requirements, the work should be shut down until the right person can get to the facility. This requires helicopter flights or boat trips, etc. It may take several days to make this happen. In the interim it is very tempting to allow someone who is not properly trained to attempt the work. This can introduce significant risk.

Discipline also comes into the process with respect to making sure the operator knows who is working on the facility. There needs to be some method to check that the specific individuals the contractor was to supply are indeed the ones who showed up on the facility. Some of these contractors are very large organizations and people become ill, take vacations, retire, etc. The operator is responsible to provide assurance that the replacement is qualified. My personal conviction to this is a result of a situation that occurred on a platform within my area of responsibility. The facility had sustained significant damage from a hurricane, and repair work was in progress. Although we specifically requested highly experienced personnel from the contractor, when the crew arrived one member told my PIC how excited he was to have his first offshore assignment. The PIC called me and asked for permission to send this individual back on the same helicopter he arrived on. While I was pleased with the reaction of my staff, not checking who the contractor sent to the heliport cost an extra flight and most importantly left us working with one less resource at a time we needed everybody we could safely have on board.

7.10 Element eight — mechanical integrity

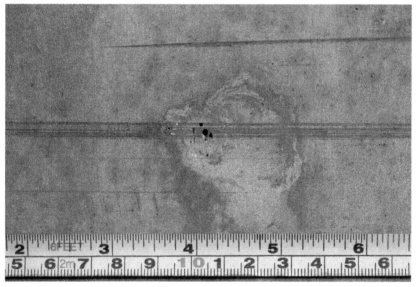

Shutterstock #1093865138

§ 250.1916 — **What criteria for mechanical integrity must my SEMS program meet?**

You must develop and implement written procedures that provide instructions to ensure the mechanical integrity and safe operation of equipment through inspection, testing, and quality assurance. The purpose of mechanical integrity is to ensure that equipment is fit for service. Your mechanical integrity program must encompass all equipment and systems used to prevent or mitigate uncontrolled releases of hydrocarbons, toxic substances, or other materials that may cause environmental or safety consequences. These procedures must address the following:

(a) The design, procurement, fabrication, installation, calibration, and maintenance of your equipment and systems in accordance with the manufacturer's design and material specifications.

(b) The training of each employee involved in maintaining your equipment and systems so that your employees can implement your mechanical integrity program.

(c) The frequency of inspections and tests of your equipment and systems.

The frequency of inspections and tests must be in accordance with BSEE regulations and meet the manufacturer's recommendations. Inspections and tests can be performed more frequently if determined to be necessary by prior operating experience.

(d) The documentation of each inspection and test that has been performed on your equipment and systems. This documentation must identify the date of the inspection or test; include the name and position, and the signature of the person who performed the inspection or test; include the serial number or other identifier of the equipment on which the inspection or test was performed; include a description of the inspection or test performed; and the results of the inspection test.

(e) The correction of deficiencies associated with equipment and systems that are outside the manufacturer's recommended limits. Such corrections must be made before further use of the equipment and system.

(f) The installation of new equipment and constructing systems. The procedures must address the application for which they will be used.

(g) The modification of existing equipment and systems. The procedures must ensure that they are modified for the application for which they will be used.

(h) The verification that inspections and tests are being performed. The procedures must be appropriate to ensure that equipment and systems are installed consistent with design specifications and the manufacturer's instructions.

(i) The assurance that maintenance materials, spare parts, and equipment are suitable for the applications for which they will be used.

7.10.1 Understanding element eight

I view mechanical integrity as the marathon of elements because it is such a long and complex element. As a SEMS audit team member I have seen this element broken into multiple sections and the associated interview taking multiple days. The Element 8 section of 30 CFR 250 subpart S is not strikingly longer than the other elements, yet the COS protocol for this element has approximately 35 questions. What makes mechanical integrity unique among the SEMS elements is the breadth of the topic. It touches essentially every part of the facility and can incorporate a wide variety of practices and requirements from API, American Society of Mechanical Engineers (ASME), American National Standards Institute (ANSI), National Association of Corrosion Engineers, (NACE), etc.

The first indication that this is a far ranging element should be the wording in 30 CFR 250.1916:

> *Your mechanical integrity program must encompass all equipment and systems used to prevent or mitigate uncontrolled releases of hydrocarbons, toxic substances or other materials that may cause environmental or safety consequences.*

This may well be different than many operators' definition of critical equipment because it does not specify the quantity of material or the severity of the release. Consequently, it is possible to develop a critical equipment listing based on corporate global guidelines and yet not be consistent with the SEMS requirements. As an audit team member, one of the first questions I ask is to see the critical equipment list and the documentation of how the list was developed. If the list is not complete or current, regardless of how thorough the rest of the program is, there is potential for risk of an uncontrolled release.

The second indication of the scope of this element is the reference to the design, procurement, fabrication, installation, testing, inspection, monitoring and maintenance of critical equipment. That pretty much covers the whole lifecycle spectrum. Add to this the fact that the operator is also responsible for the mechanical integrity of any equipment brought onto the facility by contractors or third parties. So, this element applies to any equipment that could prevent or cause an uncontrolled release and applies from the design through every step of the equipment's life. This seems pretty straightforward, right? If this seems like a difficult and daunting requirement, it is because it is. This is why the role of the SEMS program is unique with respect to mechanical integrity. The scope of this element is beyond what a SEMS coordinator, an HSSE team or an operations team can expect to understand or develop without significant SME involvement. Offshore operators with effective mechanical integrity programs generally have a significant amount of resources dedicated to mechanical integrity; in fact it represents a major part of what operators do. These facilities are installed in the harsh offshore environment and once installed the operators spend significant resources keeping it all intact and running. Think about how many offshore personnel are dedicated to maintaining the integrity of the facility in one way or another. Without mechanical integrity the facility deteriorates and the ability to produce without safety or environmental impact is lost.

So where does the management system and SEMS fit in if this is all being done already? The breadth of this topic makes it particularly susceptible to "errors of omission", or in operations speak it is prime for stuff falling through the cracks. When any aspect or part of a mechanical integrity program is not fully attended to, the results can be uncontrolled releases with all the associated safety and environmental consequences. The purpose of the management system is to provide the framework and the process to assure that all parts of the program are in place and implemented thoroughly. This may be best illustrated by looking at what it takes to effectively audit a mechanical integrity program. The associated interview sessions are usually divided into multiple specific topics such as subsea, topsides, pressure vessels, piping, structural, downhole, rotating equipment, etc. In many cases the personnel associated with each part of the program are very specialized.

Because mechanical integrity is such a broad element and takes up so much time on the SEMS audit agenda it is tempting for an organization to begin seeing the SEMS coordinator as the keeper of the mechanical integrity program as sometime the SEMS coordinator is the only person who knows how broad the mechanical integrity program is. This is neither effective nor practical. The technical assurance for the mechanical integrity program should come from subject matter experts working with operations personnel. The role of the SEMS Program and SEMS coordinator is to provide an overarching assurance that all aspects of mechanical integrity (as defined by the subject matter experts) are included in the management system. This means that all aspects of the mechanical integrity program must be fully documented in detail. The documented program must meet all internal and external requirements for each aspect of the program. For example, pressure vessel inspection and maintenance likely involves multiple API, ANSI, ASME, NACE etc. requirements. And then comes the truly hard part — you must document sufficiently to demonstrate adherence to your program. The documentation is not simply to allow the operator to successfully pass their SEMS audit. It is far more important than that. Without the documentation the operator's ability to assess the effectiveness of the program is lost, the ability to provide assurance that critical equipment is being maintained effectively is lost, and the ability to provide assurance that the equipment is suitable for the service it is in is also lost. In the event of an incident or near miss there is insufficient data to evaluate the role mechanical integrity or failure played in the event.

7.10.2 What good looks like

The best mechanical integrity programs I have reviewed are well organized and logical to follow. There is no specific formula or template for this, as the program should reflect the operator's facilities and operations. An operator with multiple deepwater platforms and drilling operations requires a much different mechanical integrity program than an operator with a few unmanned shallow water facilities. Regardless of the scope of the operations, the goal of the mechanical integrity program is the same; to prevent or mitigate uncontrolled releases of hydrocarbons, toxic substances or other materials that may cause environmental or safety consequences. This means the mechanical integrity program must cover all facilities, not just the ones that are easy to monitor. While this is not an exhaustive list, I have seen mechanical integrity programs which cover:

- Wellbores
- Subsea wellheads
- Subsea manifolds and tie ins
- Subsea pipelines
- Risers and flex joints
- Hull
- Topsides equipment
- Structure
- Drilling facilities
- Third party equipment
- Rental equipment

As I said, this is not a complete list but is intended to illustrate just how far reaching this element can be. I certainly understand that some of this equipment cannot be monitored or protected using conventional methods, as it can be thousands of feet underwater. This does not mean it can be ignored, but it does introduce limitations that must be part of the considerations when developing a mechanical integrity program.

A good mechanical integrity program is based upon a comprehensive inventory of the critical equipment, as defined by 30 CFR 250 subpart S. The inventory should be detailed and specific to the facility. In the early days of BSEE SEMS implementation, many operators designated everything as critical equipment. While this sounds good at first, this can unnecessarily complicate the program and dilute the emphasis on the truly critical equipment. There is equipment on an offshore facility that if it fails, there is no risk of any loss of containment. I did not say loss of revenue. SEMS is focused on the prevention of uncontrolled releases, not the operator's bottom line. Take time to thoroughly identify and inventory the critical equipment. This is the list of equipment that Element 8 requires the operator to address from design thorough the full lifecycle.

At this point I am going to jump to the activities most associated with operations; testing, inspection and maintenance of critical equipment. It is important to note that Element 8 does include the design, procurement, fabrication, and installation of critical equipment. If an operator is involved in fabrication of a new facility or a significant modification of an existing facility, in order to comply with BSEE SEMS they must have processes in place and be able to demonstrate that the critical equipment was designed, fabricated and installed according the specifications developed by the appropriate SMEs. I am not going to go into detail regarding these stages as my experience is that these aspects of the mechanical integrity program are usually handled by a project group or contractor which has existing systems in place for this purpose. Just remember, it is the operator's responsibility to assure these processes are in place. The exceptions to this are small modifications and spare parts. This responsibility can fall to the operations team and an assurance process must be in place, and should incorporate an effective MOC process.

It is also my experience that the most adverse findings come from the testing, inspection and maintenance phases. If you think about it, this is where there are so many opportunities for things to fall through the cracks. The requirements for each item on the critical equipment inventory can vary significantly in frequency and difficulty. Some may be completed by the operations personnel on a monthly basis while others require certified inspectors with an inspection frequency of several years. Keeping track of the requirements and assuring that they are completed requires a well organized process. This is another point where the process can be fit for purpose with respect to the operator's facilities. Multiple facilities each with long critical equipment inventories generally require investment in some type of tracking tool with scheduling and reminder functions. A single facility with a short critical equipment list requires less elaborate tools but just as much diligence.

The first step after identification of the critical equipment inventory is for the SMEs to develop the testing, inspection and maintenance requirements for each item. This is complex and can involve multiple codes, standards, regulations and manufacturer recommendations. There is no easy way or shortcut to do this. You must invest in the technical resources to get this correct, and you must document the basis for the program you develop. In the case of a good mechanical integrity program, it should be an easy request to provide an auditor or regulator the mechanical integrity requirements for any piece of critical equipment and the basis of those requirements.

The next request by an auditor or regulator will likely be evidence that the required testing, inspection and maintenance is being carried out. If the documentation required for developing the mechanical integrity program seems extensive, the documentation required to demonstrate implementation is even more. This is best illustrated utilizing the COS audit protocol question 1200:

"Do the mechanical integrity procedures address the documentation of each inspection and test that has been performed on equipment and systems?

Does the documentation identify:
- The date of the inspection or test?
- The name and position, and the signature of the person who performed the inspection or test?
- The serial number or other identifier of the equipment on which the inspection or test was performed?
- A description of the inspection or test performed?
- The results of the inspection or test?"

Obviously this requires discipline and an effective means of maintaining this information. Again, in an effective program it should be easy to produce the information regarding a previous test for an auditor, regulator, or incident investigator.

If you have experienced a SEMS audit or inspection, then you already know the next topic. If the test or inspection found that any corrective actions are required can you demonstrate how you delegate, track and assure completion of the required actions? For example, if a piping non-destructive testing (NDT) report indicates that a section of the facility's piping no longer meets the wall thickness requirements for the service it is in, an operator needs to show how the corrective action was developed, who it was assigned to and confirmation that it was completed. This is more emphasis on effective documentation.

Maintenance of critical equipment is often overlooked, if not in the completion of it then in the documentation of it. There is a lot of equipment requiring maintenance on an offshore facility. A good mechanical integrity program will include a method of prioritizing the critical equipment maintenance. Let's be honest, maintenance schedules can be delayed due to a variety of reasons; availability of crews, weather, availability of parts, etc. An offshore facility has only so many beds and a limited amount of space. The good systems prioritize getting the critical equipment done first when limitations impact the maintenance schedule. A common point of confusion relates to understanding that maintenance and inspection/testing are not the same thing. Just because a piece of equipment was tested and functioned correctly does not mean it is being properly maintained. If the manufacturer of a piece of critical equipment recommends quarterly lubrication, then the periodic confirmation that the equipment is operational does not suffice to deem it as being properly maintained. "Running to failure" is not generally a good practice for critical equipment.

During the mechanical integrity section of a SEMS audit, the SEMS coordinator role is more like an orchestra conductor than a musician, calling upon a variety of SMEs, operations personnel and document control teams.

7.11 Element nine — pre start-up review

§ 250.1917 — What criteria for pre-startup review must be in my SEMS program?

Your SEMS program must require that the commissioning process include a pre-startup safety and environmental review for new and significantly modified facilities that are subject to this subpart to confirm that the following criteria are met:

(a) Construction and equipment are in accordance with applicable specifications.

(b) Safety, environmental, operating, maintenance, and emergency procedures are in place and are adequate.

(c) Safety and environmental information is current.

(d) Hazards analysis recommendations have been implemented as appropriate.

(e) Training of operating personnel has been completed.

(f) Programs to address management of change and other elements of this subpart are in place.

(g) Safe work practices are in place.

7.11.1 Understanding element nine

In the COS SEMS audit protocol (COS-1-01) the pre start up review (PSR) element consists of one question with seven sections. And the sections ask questions that are what you would expect a comprehensive pre start-up review to address. So at first look, this appears to be a free space on the bingo card. Pull up your PSR form and associated procedure document and some completed PSRs and you should be done. Many times the PSR is an integral part of the operator's MOC process. The story is common, the operator opens up their Management of Change (MOC) program document and there is a wonderful PSR template that is required to be completed before the work associated with an MOC may be started up. They may even have multiple examples of completed PSR templates associated with MOCs.

Looking across the conference room table it is apparent the audit team is not finished with this element. The room goes silent until an audit team member speaks up and asks about the PSR process for:

- Bringing on new wells
- Bringing a new rig into the field to drill
- Commissioning new facilities
- Newly acquired operations

This generally results in taking a break while the drilling team and the project teams are contacted and asked to come to the conference room, to speak about a topic they were not even aware of. Many times there is indeed a documented process that satisfies these requirements and after finding the right personnel to speak to it the audit continues.

The learning here is to be aware that pre start-up review is a term that many operations personnel closely associate with MOCs. However, the concept of PSR covers a much wider range of activity, and if you are focusing your SEMS program on reducing risk, consideration of all avenues where there are new operations starting up is important.

7.11.2 **What good looks like**

To do PSR well, it is simply a matter of awareness and discipline. Awareness in that many people within the operator and contractor organizations need to know that the PSR step is required and when it is required. Those organizations that do it well seem to have it as an integral part of their work processes and err on the side of doing more PSRs rather than less. In fact, when you consider the 30 CFR 250 Subpart S wording, it is hard to imagine that there are many cases where an MOC is required but a PSR is not. I agree, there are exceptions but many operators consider this when determining the need for a PSR. As an audit team member, I consider an MOC that does not have an associated PSR worthy of further investigation.

As discussed above, not all PSR implementations are tied to an MOC. I have seen a variety of documents which are essentially pre start up reviews but are called something else:

- Well handover process from drilling to operations
- Project handover process from the project team to operations
- Drilling rig vetting procedures
- Acquisition handover to operations
- Etc.

The awareness carries over to the physical start up operations. The operations leadership knows not to start up anything without a completed PSR, or the completion of one of the other pre start up review processes. I have observed an OIM holding up the start up of a significant facility expansion while he waited for someone to show him a completed PSR. Some of the best PSR processes I have witnessed separated the approval of the PSR and the approval to begin start up activities, leaving the latter responsibility to someone on the facility with knowledge of the ongoing activity and ambient conditions on the facility.

The discipline comes in the assurance that a PSR is completed every time it is required, because once something is allowed to start up or be changed without going through the PSR process it becomes very easy to do it again. This is difficult when the modification may appear relatively minor. Take a small piping change that everyone on a small facility watched be completed and is aware of the change. This seems pretty benign at first. However, without a PSR there is no assurance that revised drawings, specifications operating procedures, etc. were developed and the files updated. Fast forward a few years and the original crew are gone or the facility is under new operatorship. The new operators could follow an incorrect procedure or base a decision on an incorrect drawing.

7.12 Element ten — emergency response and control

Shutterstock # 152407577

§ 250.1918 — **What criteria for emergency response and control must be in my SEMS program?**

Your SEMS program must require that emergency response and control plans are in place and are ready for immediate implementation. These plans must be validated by drills carried out in accordance with a schedule defined by the SEMS training program (§ 250.1915). The SEMS emergency response and control plans must include:

(a) Emergency Action Plan that assigns authority and responsibility to the appropriate qualified person(s) at a facility for initiating effective emergency response and control, addressing emergency reporting and response requirements, and complying with all applicable governmental regulations;

(b) Emergency Control Center(s) designated for each facility with access to the Emergency Action Plans, oil spill contingency plan, and other safety and environmental information (§ 250.1910); and

(c) Training and Drills incorporating emergency response and evacuation procedures conducted periodically for all personnel (including contractor's personnel), as required by the SEMS training program (§ 250.1915). Drills must be based on realistic scenarios conducted periodically to exercise elements contained in the facility or area emergency action plan. An analysis and critique of each drill must be conducted to identify and correct weaknesses.

7.12.1 Understanding element ten

This element generally has two or more documents associated with it. First is a BSEE approved Oil Spill Response Plan (OSRP). The Federal Water Pollution Control Act requires that owners or operators of offshore facilities and associated pipelines prepare and submit OSRPs to BSEE. The requirements for the OSRP are found in 30 CFR Part 254, "Oil-Spill Response Requirements for Facilities Located Seaward of the Coastline". OSRPs are large documents, involving hundreds of pages. Before you let this discourage you, there are many companies who are well versed in the preparation of these documents as well as the implementation requirements. There is no reason you need to become an expert on developing these plans, just focus on the implementation.

The next document can be a single document or multiple documents, it does not matter. What matters is that the document(s) address emergency response and control in relation to:
- Fire and/or blowouts
- Collision
- Spills of hazardous substances
- Emergency evacuation procedures

You may find yourself looking at the wording from CFR 250 Subpart S included above and you don't see these listed there. They are not there. However, they are a part of the COS audit protocol (COS-1-01). And while emergency evacuation procedures are also a part of Element 5, Operating Procedures, I include it here as many times emergency response procedures are integral with emergency response documents. The documentation requirement is further expanded if the facility is a floating facility. Floating facilities require documentation consistent with the US Coast Guard requirements and include such items as a current Marine Operating Manual (MOM). The key here is to take the time to enlist the help of the appropriate SMEs to develop a comprehensive listing of the emergency response and control procedures and documents that your specific facility is required to have. This is important so take the extra time and resources to assure you have all the procedures in place and current.

Everybody thinks they have this element handled. They have annual OSRP drills, they have monthly drills for the other scenarios, they have emergency response plans documented and in place. And generally, offshore operators pay a lot of attention to this, not only because of the multiple associated regulations but because emergencies offshore bring with them added complexity and hazards. As one driller once told me "there ain't no place to run". Yet I have observed gaps within the programs of conscientious offshore operators. How does this happen? My opinion is that some programs have been in place so long they are on "autopilot" and have not been updated per best practices. There are three areas where gaps most frequently show up.

First, remember Element 2, Safety and Environmental Information? Included in the information that SEMS requires an operator to maintain is information that may be required to respond to an emergency. This means that personnel on the facility must have access to this information as well as those on the beach, including the incident command center which may or may not be located at the operator's office. This means the current, updated information which is generally in electronic format. This also means the response time of the information systems offshore must be sufficiently fast to support emergency operations or there must be a controlled hard copy process in place. Using outdated information or waiting for slow downloads is not acceptable for emergency operations.

Second, make sure your emergency action plans cover all the situations required by SEMS. I have observed multiple operators that have fires, spills, and blowouts covered, but the COS audit protocol specifically asks for an emergency action plan for collisions. I have no explanation for why this happens, just that I have observed it. One operator did tell me that a collision was likely to result in a spill or fire so they did not see the need for a collision plan. Just prepare a collision emergency action plan. Similarly, if you have made the decision that you will not have on board personnel attempt to put out fires but rather to immediately evacuate in case of fire, say that clearly in the documentation.

Third, a drill is not truly a drill unless you take the time to critique the drill, document the critique and make recommendations for improvements. And recommendations are not recommendations unless they are assigned to someone and progress is tracked. While this may not sound like a big deal, on a large facility with multiple hitches and multiple shifts, drills for all personnel once a month can generate a lot of drill records and recommendations to maintain and track.

There is a situation that is somewhat unique to the offshore industry that warrants mention, which is the utilization of a mobile offshore drilling unit (MODU). Generally the MODU is owned and operated by a drilling contractor, but the operator may have personnel on board. If an emergency occurs on the MODU, the roles of the drilling contractor and the operator must be clear and consistent. This may seem simple, if it happens on the MODU the drilling contractor responds to the emergency. This can be more complicated with respect to spills and releases. Many operators want to "own" the response to releases and/or spills. Have this clarified before there is oil on the water. A good practice in these circumstances is to conduct a drill where both the drilling contractor and the operator mobilize their respective emergency procedures and control centers in an effort to simulate the cooperation and communications needed in what is a complicated situation.

7.12.2 What good looks like

Emergency response and control is, in my opinion, one of the easiest elements to audit and one of the easiest to describe what good looks like. Much of the section above on understanding this element focused on the documentation. It is a relatively straightforward exercise to review the documentation and determine that it does or does not meet the requirements. The more difficult aspect to evaluate is the awareness and understanding of the plans by the involved personnel. This involves both onshore and offshore personnel.

All personnel who have a role in any of the emergency response procedures should be acutely aware of their role and be comfortable that they are capable of doing everything they may be called upon to do. For example, if a person has the delegation as the incident commander that person should have all the required training and have drilled in that role. The worst time to be figuring out what your role is and what it entails is during a real incident. As an auditor I ask the office personnel where the command center is, what their roles are and when they last drilled. During the offshore portion of an audit I ask the same questions of the operations personnel as well as ask them to locate the S&EI information they may be required to utilize.

If you truly want to assess organizations' preparedness for emergency response, observe a drill. Prior to becoming a SEMS auditor, I was involved with a lot of drills, both as a participant and as a leadership assessor. While drills can take anywhere from hours to days, the organization's preparedness shows relatively early. Personnel know their roles and they smoothly transition from operations to emergency response mode. The command center is activated quickly and the resources needed are available. Personnel are serious about the quality of the drill and the response, and take an active part in the post drill critique. Conversely, if the first thing the personnel do is to look at some kind of duty roster or notebook to determine their role in the drill, the cabinet with the response plans and materials is empty, and there is chaos in the command center, don't expect a smooth drill.

I worked with an offshore assessor who had 40 + years of offshore operations leadership experience and prior to the time the drill was to begin he would quietly go back to his room while I went to the command center on the facility. On one occasion the facility completed the drill, which included a muster for evacuation and this individual was never missed or looked for. This individual then proudly walked into the drill critique meeting and announced he had been left behind. This trick only works a couple of times — the platforms did tell each other of his practice!

7.13 Element eleven — investigation of incidents

§250.1919 – What criteria for investigation of incidents must be in my SEMS program?

To learn from incidents and help prevent similar incidents, your SEMS program must establish procedures for investigation of all incidents with serious safety or environmental consequences and require investigation of incidents that are determined by facility management or BSEE to have possessed the potential for serious safety or environmental consequences. Incident investigations must be initiated as promptly as possible, with due regard for the necessity of securing the incident scene and protecting people and the environment. Incident investigations must be conducted by personnel knowledgeable in the process involved, investigation techniques, and other specialties that are relevant or necessary.

(a) The investigation of an incident must address the following:

(1) The nature of the incident;

(2) The factors (human or other) that contributed to the initiation of the incident and its escalation/control; and

(3) Recommended changes identified as a result of the investigation.

(b) A corrective action program must be established based on the findings of the investigation in order to analyze incidents for common root causes. The corrective action program must:

(1) Retain the findings of investigations for use in the next hazard analysis update or audit;

(2) Determine and document the response to each finding to ensure that corrective actions are completed; and

(3) Implement a system whereby conclusions of investigations are distributed to similar facilities and appropriate personnel within their organization.

7.13.1 Understanding element eleven

It is simple; when there is an incident you should investigate it. Those who investigate it should be qualified and knowledgeable in the operations, and the person facilitating the investigation should be trained in some type of root cause analysis. Any actions required as a result of the investigation should be assigned to someone and the progress tracked regularly as a performance metric. If it is this simple, why are there still gaps in operators' incident investigation procedures?

At the core of any incident investigation program should be a thorough document that details the operator's incident investigation process and procedures. The purpose of that document is not only to present how incident investigations are to be conducted, but to assure that incidents of similar consequence are investigated consistently. There are many methods of incident investigation and root cause determination, and the time to decide which method your organization will use is not at the time of the incident. Neither is it the time to decide what the makeup of the investigation team should look like. Too many times I have witnessed incident investigations that were completed by personnel who were simply those accessible at the time of the incident and the investigation method was whatever this group was familiar with. A thorough incident investigation process document does not eliminate the need for qualified personnel to make decisions and recommendations regarding the investigation, but the document should guide most of the decisions and process.

This can be illustrated by a very personal example. I had traveled from Houston to Chicago on a Sunday and was looking forward to attending a companywide symposium on operations excellence. I had even read the pre read and was expecting some timely learning. By 10:00 a.m. on Monday I was tapped on the shoulder by a nice lady from the hotel staff who handed me a note to please leave the room immediately and call an operations vice president for an asset that I was not responsible for. Wondering what was so important he could not leave a message on my mobile; I gracefully exited the full room and called. He immediately picked up the phone and informed me I had a 5:00 p.m. incident investigation meeting to lead in Houston, and that my travel arrangements were made and he hoped I did not mind that the hotel staffed had repacked my suitcase and it was in the cab in front of the hotel. There had been a major incident on a large deepwater facility that fortunately had not injured anyone but had caused significant damage and in slightly different circumstances had the potential for fatalities. The organization's incident investigation process documents required the investigation be led by an individual external to the leadership of the affected facility, trained in severe investigation facilitation, and in a leadership role in another part of the organization. The document further specified the skill sets and experience of the members of the investigation team and the investigation facilitator. We met at 5:00 p.m. as scheduled and found ourselves headed offshore early the next morning.

If you have come this far in this book, the next reminder is going to be familiar. Incident investigations must be documented. While the reference materials do not recommend a specific template or form, 30 CFR

250 Subpart S is relatively specific about what needs to be recorded. Reading the specific language included at the start of this section it is pretty clear you need to record:

- The nature of the incident
- The incident investigation team members
- The factors contributing to the incident
- Recommended changes

It is also required that you retain the findings for use in the next hazard analysis or audit.

The documented process should also provide guidance regarding the desired timing to complete investigations. This can be a bit of a controversial point, as time is of the essence when completing an investigation (as witnessed by my 36 hours Chicago to GOM experience), but there are times when critical information takes significant time to gather. For example, if metallurgical analysis of a failed part is required, this can take several weeks and maybe longer to get complete results. Suffice to say, the sooner the better and in the interest of avoiding similar incidents investigations are high priority.

Another potential area of confusion regards the SEMS requirement to include the investigation of incidents that had the potential for serious consequence. This does not mean simply tracking near misses; it means investigating those that could have been serious in the same manner a serious incident would have been investigated. This is not incident "investigation light". This is the full meal deal, complete with write up and action items.

7.13.2 What good looks like

This element is relatively easy to audit. In general I go straight to the incident investigation reports, even before reviewing the incident investigation program documents. Similar to management of change documents, a good incident report is one stop shopping. An auditor should be able to read an incident report and not require any other documentation to fully understand:

- What actually happened and in detail. A one line of "fire in the treater" is not sufficient. The report should have all the details as developed by the investigation team, and in chronological order.
- The members of the investigation team, their qualifications and how they meet the requirements as set out in the incident investigation process document

- The incident investigation process used and how it met the requirements as set out in the incident investigation process document
- The investigation team conclusion regarding the root cause(s) of the incident
- Recommended actions to prevent recurrence
- Communications plan to share results with other facilities
- Required agency notification records and communications
- Record of review and approval of the investigation by the appropriate level of management.

I have reviewed incident investigation reports that included more than this, but this is my personal checklist for the basics of a "best practice" investigation report. I also expect to see some kind of method of delegation and tracking of the recommended actions, however this may be in the organizations action tracker system or tool and not in the report.

The next big clue to the thoroughness of an organizations incident investigation implementation lies in a review of the determination of the cause. There are a couple root cause phrases that are a red flag to me. The first is "operator error". While this can be a valid cause, seeing it repeatedly can be a sign that the investigation was hurried to completion. By taking this just a couple steps further, more clarity can be gained. Specifically:

- Was the operator error due to insufficient training?
- Was the operator error a result of external pressures such as a deadline?
- Was the operator fatigued?
- Etc.

When a well trained, competent individual makes a mistake it is worth a few more questions to dig a bit deeper.

Similarly, "equipment failure" is another red flag for me. Yes, equipment fails offshore. Operators have maintenance programs, secondary containment, safety shut downs, level alarms, just to name a few of the ways we work to prevent equipment failure from resulting in an incident. As with operator error I recommend that when equipment failure results in an incident investigation, ask a few more questions:

- Was the failure related to maintenance?
- Was the failure a result of safety equipment malfunction?
- Was the failure a result of design issues?
- Was the failure a result of changing conditions?
- Etc.

Simply replacing the failed equipment may not be an effective method to prevent future incidents.

7.14 Element twelve — auditing

Shutterstock #372560002.

There are three sections of CFR 250 Subpart S that are associated with the audit process. The first (§ 250.1920) relates to the actual audit requirements, the second (§ 250.1921) relates to the requirements of the Audit Service Provider (ASP) and the third (§ 250.1920) relates to the Accreditation Body (AB) which accredited the ASP.

(§ 250.1920) — What are the auditing requirements for my SEMS program?

(a) Your SEMS program must be audited by an accredited ASP according to the requirements of this subpart and API RP 75, Section 12 (incorporated by reference as specified in § 250.198). The audit process must also meet or exceed the criteria in Sections 9.1 through 9.8 of *Requirements for Third-party SEMS Auditing and Certification of Deepwater Operations* COS−2−03 (incorporated by reference as specified in § 250.198) or its equivalent. Additionally, the audit team lead must be an employee, representative, or agent of the ASP, and must not have any affiliation with the operator. The remaining team members may be chosen from your personnel and those of the ASP. The audit must be comprehensive and include all elements of your SEMS program. It must also identify safety and environmental performance deficiencies.

(b) Your audit plan and procedures must meet or exceed all of the recommendations included in API RP 75 Section 12 (as specified in § 250.198) and include information on how you addressed those recommendations. You must specifically address the following items:

(1) Section 12.1 General.

(2) Section 12.2 Scope.

(3) Section 12.3 Audit Coverage.

(4) Section 12.4 Audit Plan. You must submit your written Audit Plan to BSEE reserves the right to modify the list of facilities that you propose to audit.

(5) Section 12.5 Audit Frequency, except your audit interval, must not exceed 3 years after the 2-year time period for the first audit. The 3-year auditing cycle begins on the start date of each comprehensive audit (including the initial implementation audit) and ends on the start date of your next comprehensive audit.

(6) Section 12.6 Audit Team. Your audits must be performed by an ASP as described in § 250.1921. You must include the ASPs qualifications in your audit plan.

(c) You must submit an audit report of the audit findings, observations, deficiencies identified, and conclusions to BSEE within 60 days of the audit completion date.

(d) You must provide BSEE with a copy of your CAP for addressing the deficiencies identified in your audit within 60 days of the audit completion date. Your CAP must include the name and job title of the personnel responsible for correcting the identified deficiency(ies). The BSEE will notify you as soon as practicable after receipt of your CAP if your proposed schedule is not acceptable or if the CAP does not effectively address the audit findings.

§ 250.1921 – What qualifications must the ASP meet?

(a) The ASP must meet or exceed the qualifications, competency, and training criteria contained in Section 3 and Sections 6 through 10 of *Qualification and Competence Requirements for Audit Teams and Auditors Performing Third party SEMS Audits of Deepwater Operations*, COS–2–01, (incorporated by reference as specified in § 250.198) or its equivalent;

(b) The ASP must be accredited by a BSEE-approved AB; and

(c) The ASP must perform an audit in accordance with 250.1920(a).

§ 250.1922 – What qualifications must an AB meet?

(a) In order for BSEE to approve an AB, the organization must satisfy the requirements of the International Organization for Standardization's (ISO/ IEC 17011) *Conformity assessment—General requirements for accreditation bodies accrediting conformity assessment bodies*, First Edition 2004−09−01; Corrected Version 2005−02−15 (incorporated by reference as specified in § 250.198) or its equivalent.

(1) The AB must have an accreditation process that meets or exceeds the requirements contained in Section 6 of *Requirements for Accreditation of Audit Service Providers Performing SEMS Audits and Certification of Deepwater Operations*, COS−2−04 (incorporated by reference as specified in § 250.198) or its equivalent, and other requirements specified in this subpart. Organizations requesting approval must submit documentation to BSEE describing the process for assessing an ASP for accreditation and approving, maintaining, and withdrawing the accreditation of an ASP. Requests for approval must be sent to DOI/BSEE, ATTN: Chief, Office of Offshore Regulatory Programs, 381 Elden Street, HE−3314, Herndon, VA 20170.

(2) An AB may be subject to BSEE audits and other requirements deemed necessary to verify compliance with the accreditation requirements.

(b) An AB must have procedures in place to avoid conflicts of interest with the ASP and make such information available to BSEE upon request.

7.14.1 Understanding element twelve

This element should be the one that requires the least amount of interview time during the third party SEMS audits because of all the elements this one may have the least interpretation and "grey area". BSEE set out the process to be followed which includes:

- How often audits are conducted
- What percentage of the operations must be audited
- Process for getting your audit plan approved
- Audit results reporting requirement
- Corrective Action Plan (CAP) requirements

- Approved Audit Service Providers (ASP)
- Qualifications for audit team members

If an operator follows the BSEE instructions, this element can be a relatively easy one to demonstrate compliance with. So on to Element 13? Not so fast. There is one aspect of Element 12 that can cause issues that ripple into multiple other elements.

The corrective action plan (CAP). Even the name implies that something needs to be done to correct something not being consistent with the SEMS requirements. Developing the CAP and sending it to BSEE is just the start of the process. Reporting your progress to BSEE as required is also a part of the process. The big part is actually doing the work required to correct the finding. The corrective action needs to be assigned to the appropriate personnel, the action plan must be developed in detail, the action plan must be implemented, and the operator must be able to demonstrate that the change has been made and is fully implemented across the facility.

Generally, this demonstration involves the use of the MOC process (Element 4). The process should include (but is not limited to):

- Revision or creation of documents
- Installation of new equipment
- Changes in procedures
- Changes in process or operating limits
- Hazard analysis
- Approval by the appropriate level of delegation
- Training of impacted personnel
- The PSR
- MOC approval for closure

If any part of this is missing, the change has not been implemented. This means any risk that was to be mitigated by the action may remain in the system. To top it off, audit teams are required to look for verification of the implementation of the CAP from previous audits. Put yourself in BSEEs position for a moment. Three years ago the third party audit identified a gap that required actions of some type to bring the operator's SEMS Program into compliance. The operator has sent in progress updates that show the action as completed. Yet when the next audit is completed the same gap still exists because the required changes were not fully implemented. As an audit team member this casts doubt regarding all actions from previous audits.

7.14.2 What good looks like

As I mentioned earlier in this section, 30 CFR 250 Subpart S provides significant detail regarding the periodic third party SEMS audit requirements. By this point in the implementation of SEMS in GOM, it is very rare to find an operator who is not aware of and following those requirements with respect to completing the audits. What does separate those who do this well and those who do not is the completion of the CAP, most of which is discussed in "Understanding Element 12".

For those organizations that do this well this element is a "non event", and truly is that free square on the bingo card. Such operators are able to provide the audit team the CAP from the last audit, the periodic reports of progress to BSEE and evidence that the action items were indeed brought to closure. Usually this is in the form of MOCs but can take other forms. However, I have seen a surprising variety and creativity in the ways operators can fail to complete actions on their CAP. This is a problem for several reasons. First, the risk associated with the action item has not been mitigated. Second, when the next audit is done the audit team most likely will uncover this and it will be in the audit report as a repeat finding. BSEE truly sees SEMS as a continuous improvement process, and repeat findings are not an indication of improvement efforts.

I have observed operators who simply present the letter from BSEE indicating that BSEE received their final CAP status report as the documentation for this element and are ready to move on. They are then disappointed when the audit team wants more details. BSEE cannot possibly understand the operations of every operator in the OCS, remember the details of every finding of every audit, and provide assurance that every action item was completed. However, the ASP may decide to spend time reviewing the completion of the previous CAP.

Do not change the CAP after submittal to BSEE without authorization from BSEE. For example, if after further thought, an operator decides that they don't agree with a finding do not simply close the associated action out without completion of the activities as originally submitted to BSEE. If you are going to depart from the CAP as provided to BSEE, make sure to get BSEE approval prior to changing the CAP.

7.15 Element thirteen — recordkeeping

Shutterstock #246719938

§ 250.1928 — **What are my recordkeeping and documentation requirements?**

(a) Your SEMS program procedures must ensure that records and documents are maintained for a period of 6 years, except as provided below. You must document and keep all SEMS audits for 6 years and make them available to BSEE upon request. You must maintain a copy of all SEMS program documents at an onshore location.

(b) For JSAs, the person in charge of the job must document the results of the JSA in writing and must ensure that records are kept onsite for 30 days. In the case of a MODU, records must be kept onsite for 30 days or until you release the MODU, whichever comes first. You must retain these records for 2 years and make them available to BSEE upon request.

(c) You must document and date all management of change provisions as specified in § 250.1912. You must retain these records for 2 years and make them available to BSEE upon request.

(d) You must keep your injury/illness log for 2 years and make them available to BSEE upon request.

(e) You must keep all evaluations completed on contractor's safety policies and procedures for 2 years and make them available to BSEE upon request.

(f) For SWA, you must document all training and reviews required by § 250.1930(e). You must ensure that these records are kept onsite for 30 days. In the case of a MODU, records must be kept onsite for 30 days or until you release the MODU, whichever comes first. You must retain these records for 2 years and make them available to BSEE upon request.

(g) For EPP, you must document your employees' participation in the development and implementation of the SEMS program. You must retain these records for 2 years and make them available to BSEE upon request. (h) You must keep all records in an orderly manner, readily identifiable, retrievable and legible, and include the date of any and all revisions.

§ 250.1929 – What are my responsibilities for submitting OCS performance measure data?

You must submit Form BSEE–0131 on an annual basis by March 31st. The form must be broken down quarterly, reporting the previous calendar year's data.

7.15.1 Understanding element thirteen

There is no way any operator should ever have a finding in this element! The requirements are crystal clear and not subject to interpretation. They tell you what documents to keep/maintain and how long to keep/maintain them. All the operator needs to do is to assure that it is someone's responsibility to make sure this happens, and periodically check to see that it is being done.

BSEE does not tell you how to save the documents, so don't get a finding because you are debating internally how the documents are to be saved. For example, SEMS requires 30 days of JSAs to be kept on site and two years of JSAs to be available upon request. I have seen situations where the office staff is developing a process to scan and save the JSAs electronically, deciding which team will be responsible to scan the documents, which team is responsible for maintaining the electronic data base, etc. While this is going on, nobody has told the field to save the JSAs. This becomes a finding that could have been prevented by just putting the JSAs in a box. This is not only an audit issue, as any incident investigation will also ask for the JSAs. Step one should be to assure the documents are being maintained; developing the long term processes and procedures to streamline the process can come later.

A related issue that operators need to consider but that is not specifically called out in the requirements is that of document control. Do not go to the effort to develop process documents that comply with SEMS or electronic data bases only to store them in a mode where personnel unfamiliar with the requirements of SEMS can alter or destroy the records.

A quick note regarding the submittal of the BSEE-0131 form as required by § 250.1929. This should be even easier than the recordkeeping defined by § 250.1928 with respect to compliance. BSEE-0131 is a simple form which requires the operator to submit the following information broken out for production operations, drilling operations and construction operations:

- No. of Company Employee Recordable Injuries/Illnesses
- No. of Contract Employee Recordable Injuries/Illnesses
- No. of Company Employee DART Injuries/Illnesses
- No. of Contract Employee DART Injuries/Illnesses
- Company Employee Hours Worked
- Contract Employee Hours Worked
- No. of EPA NPDES Noncompliances
- No. of Spills < 1bbl
- Total Volume for Spills <1bbl

Go to the BSEE website, get the form and the instructions and get it submitted on time with your data from the previous year. Enough said.

7.15.2 What good looks like

This element should truly be another non event during the audit. In fact, the office portion of this element can be handled as pre read if you are willing to give the audit team access to your electronic data bases. This varies between organizations and individual comfort levels, but as an audit team member I find it very efficient to be able to look through the required records before arriving at the operator's office for the interviews.

With respect to the records that are required to be maintained offshore on the facility, remember that these records need to be easily accessed for auditors or BSEE personnel. The personnel on the facility should know where the records are, and what the requirements are for the maintenance of records on site. For those records such as JSAs where 30 days of records must be kept on the facility, make sure the operations personnel know that means the actual hard copies. Some of you will think this sounds a bit ridiculous in the digital world we live in, but I have witnessed the situation where the personnel on the facility were scanning the JSAs in on a daily basis and somebody decided if they were scanned it was fine to dispose of the hard copies. During the audit it was discovered that the settings on the scanner were incorrect and the JSA scans were incomplete. Just keep the paper copies.

7.16 Element fourteen — stop work authority

Shutterstock #134595269

§250.1930 — What must be included in my SEMS program for SWA?

(a) Your SWA procedures must ensure the capability to immediately stop work that is creating imminent risk or danger. These procedures must grant all personnel the responsibility and authority, without fear of reprisal, to stop work or decline to perform an assigned task when an imminent risk or danger exists. Imminent risk or danger means any condition, activity, or practice in the workplace that could reasonably be expected to cause:

(1) Death or serious physical harm; or

(2) Significant environmental harm to:

   (i) Land;

   (ii) Air; or

   (iii) Mineral deposits, marine, coastal, or human environment.

(b) The person in charge of the conducted work is responsible for ensuring the work is stopped in an orderly and safe manner. Individuals who receive a notification to stop work must comply with that direction immediately.

(c) Work may be resumed when the individual on the facility with UWA determines that the imminent risk or danger does not exist or no longer exists. The decision to resume activities must be documented in writing as soon as practicable.

(d) You must include SWA procedures and expectations as a standard statement in all JSAs.

(e) You must conduct training on your SWA procedures as part of orientations for all new personnel who perform activities on the OCS. Additionally, the SWA procedures must be reviewed during all meetings focusing on safety on facilities subject to this subpart.

7.16.1 Understanding element fourteen

This is the simple sounding, much talked about topic that is so very hard to implement. Stop the work programs have been the topic of countless discussions, workshops, publications and operators have tried a myriad of creative methods of implementation. Before jumping into why this is such a unique element, one needs to understand what the basic SEMS requirements are. The SEMS requirements apply to any work that is creating imminent risk or danger, which is further defined as any condition, practice or activity that could be expected to cause; death or serious physical harm, or significant environmental harm. An operator's SEMS Program must include a Stop Work Authority (SWA) procedure with some of the more significant requirements being:

- Granting all personnel the right to stop work that is creating an imminent risk or danger without fear of reprisal
- Requires the person in charge of the work to safely shut down the work

- Requires approval of the Ultimate Work Authority (UWA) to resume work
- Requires documentation of the decision to resume work
- Requires training regarding the SWA procedure as a part of all new personnel orientations
- All JSA forms include a standard statement regarding SWA procedures and expectations

Anyone who has had any involvement with SWA knows that developing the procedure document and putting the correct wording on the JSA template is the easy part. Consequently, don't get an adverse finding by not meeting these basic requirements. Develop a written SWA process that meets the requirements and include it in your SEMS program. Include SWA in your required training and orientation of all new personnel, including every contractor. Include the appropriate SWA wording on the JSA form you are using and if you decide during the bridging process (Section 7.8 Element 6, Contractor Management) to use a contractor's JSA form make sure it too has the required wording. In either case the wording must include a clear phrase assuring the reader that stopping the job has no risk or fear of reprisal. This may sound picky, but it is not only required by 30 CFR 250 Subpart S, but not having such wording may send a subtle, negative message to personnel. Finally, implement your SWA process fully, and document the specific SWA events as required by 30 CFR 250 Subpart S.

7.16.2 What good looks like

This is a difficult section to write as in my opinion it is possible to meet the majority of the SEMS requirements and not have a truly effective SWA program. Allow me to opine a bit based upon my personal experiences. In the GOM, there is a historical, cultural issue that needs to be overcome in order for any SWA process to work. I have had discussions with facility personnel regarding this topic, and in past generations, not all that long ago, stopping the job or refusing to work in a situation could get you "run off". Nobody wants to be the "guy who shut us in". As an operations manager I have had this discussion with many employees as I worked on implementation of stop the work programs. Consequently it takes consistent, constant re-enforcement from operations leadership that this is truly a valued behavior. And it only takes one instance of leadership not supporting someone who stopped the work to send the program back to square one. What good looks like begins with a sincerely dedicated leadership team who makes the effort to re-enforce this behavior.

It is important to interject some real world observations and practicality at this point. In reality, an operator does not want personnel debating in the middle of the work if the potential issue meets the SEMS description of imminent risk or danger. The best SWA procedures I have observed focus on personnel stopping the work any time they are uncomfortable, regardless of whether it ends up meeting the SEMS definition or not. These procedures include a process for assessing the risk after the job is stopped. If the situation meets the SEMS BSEE definition of being a condition, practice or activity that could be expected to cause; death or serious physical harm, or significant environmental harm then the requirements of 30 CFR 250 Subpart S are followed. If it does not, there are procedures to follow for those "lesser" situations as well.

The final evaluation of a SWA program begins when you first arrive on the facility. The required orientation should include a discussion of the SWA program. Not just a quick, check the box on the orientation form but a sincere discussion with a member of the platform leadership that this is a preferred behavior. My best example was on a large facility where members of the onsite leadership team took turns giving the orientation every time a helicopter arrived. However, the OIM also had a few minutes at the end of the orientation to stress key items, one of which was SWA. He took the time from a busy schedule every time a helicopter landed to do this.

Assuming all the previous components of the SWA program look good, the next step is to assess the "buy in" of the offshore personnel. This is difficult when you are a part of the leadership team and incredibly difficult when you are working under the constraints of an audit. I think this requires dedicated leadership and frequent personal interactions and a culture where employees truly believe in the sincerity of leadership.

7.17 Element fifteen – ultimate work authority

§ 250.1931 What must be included in my SEMS program for UWA?

(a) Your SEMS program must have a process to identify the individual with the UWA on your facility(ies). You must designate this individual taking into account all applicable USCG regulations that deal with designating a person in charge of an OCS facility. Your SEMS program must clearly define who is in charge at all times. In the event that multiple facilities, including a MODU, are attached and working together or in close proximity to one another to perform an OCS operation, your SEMS program must identify the individual with the UWA over the entire operation, including all facilities.

(b) You must ensure that all personnel clearly know who has UWA and who is in charge of a specific operation or activity at all times, including when that responsibility shifts to a different individual.

(c) The SEMS program must provide that if an emergency occurs that creates an imminent risk or danger to the health or safety of an individual, the public, or to the environment (as specified in § 250.1930(a)), the individual with the UWA is authorized to pursue the most effective action necessary in that individual's judgment for mitigating and abating the conditions or practices causing the emergency.

7.17.1 Understanding element fifteen

Do not let the relatively short sections of the regulations regarding Ultimate Work Authority (UWA) fool you into thinking that this is a simple element to implement. People often quickly reply that it is the OIM or PIC and think the review is over. Depending upon the operation, this actually may be the end of the discussion. Generally, if we are reviewing a production facility with no external operations ongoing, the OIM or PIC is indeed the UWA. However, add some workboats, remotely operated vehicle (ROV) operations, etc. into the picture and it can be more complicated. For example, when a production facility has an ROV boat working adjacent to it, you now need to consider the ROV boat Captain and crew. Many operators will consider the proximity of the ROV vessel to the operations facility, the type of work being done, and the type of incident involved. If the ROV vessel is outside a pre-determined radius the ROV vessel might operate essentially independently, including specifying that the vessel has an independent UWA. If the ROV vessel is in inside the pre-determined radius, then an agreement between the vessel Captain and the production facility OIM/PIC may be required as the ROV vessel may present a risk to the facility. It is similarly complicated on a MODU. It is likely that onboard the MODU is a leadership organization, a Captain, and various representatives of the organization they are drilling for. It is also possible that the UWA can vary due to the issue being addressed. It is critical that the

UWA delegation among these personnel is clear and documented prior to any work being initiated, and effectively communicated to all personnel on the facility.

7.17.2 What good looks like

As discussed above, this can be an element that can be relatively easy to do well or it can be very complicated, as it is very specific to the complexity of the operations being managed or audited. The first thing I look for when I arrive on a facility is some indication of who the UWA is. Generally there is a sign or some indication of this in the first room you enter after boarding the facility. Next, it should be included as a topic in the orientation, and the more complex the situation the more time needs to be spent on UWA. Organizations who do this well recognize that the information required for specific roles on the facility are not equal, and may have multiple levels of UWA discussions as a part of orientation. For a maintenance crew it may be sufficient to know who is the current UWA and how they will be notified should this change. For the leadership team of a MODU there will need to be significant time spent making sure all parties are comfortable with the details of the delegation of the UWA in a variety of situations.

After the orientation, as an audit team member I usually just ask somebody on the facility who the UWA is, and listen to the reply. The person is not expected to recite the UWA program document but needs to be fully comfortable with where and how they keep themselves aware of this. It is the responsibility of operations leadership to make this easy for the offshore personnel, not the responsibility of the offshore personnel to track down the information in documents or procedures.

The requirements of this element are more demanding when discussing the leadership roles of offshore facilities. Again, on a simple production facility the OIM/PIC is generally the UWA with some type of process in place to notify personnel if that changes. In those cases where there are operations adjacent to or impacting a production facility or in the case of a MODU the leadership needs to take time to assure that they are all comfortable with the UWA process. Where I have observed this done well the leadership teams of all organizations impacted by the UWA program and determination had spent significant time assuring agreement and understanding.

7.18 Element sixteen — employee participation plan

Shutterstock #352856471

§ 250.1932 — What are my EPP requirements?

(a) Your management must consult with their employees on the development, implementation, and modification of your SEMS program.

(b) Your management must develop a written plan of action regarding how your appropriate employees, in both your offices and those working on offshore facilities, will participate in your SEMS program development and implementation.

(c) Your management must ensure that employees have access to sections of your SEMS program that are relevant to their jobs.

7.18.1 Understanding element sixteen

It is relatively simple to develop an Employee Participation Plan (EPP) procedure, but it is difficult to obtain open and candid feedback from personnel regarding the SEMS Program. This is closely related to the much larger subject of worker empowerment which has been the subject of many studies and investigations, and is a topic that is evolving in real time. I am going to editorialize a bit here. Effectively empowering the workforce may well be the key to the next step change in offshore operations HSE performance.

I have observed a number of methods associated with this element. I can honestly say that I have yet to see one that I would put forth as a best practice. I do have examples of practices that historically have not worked well which can hopefully save you some time and resources. Emailing the SEMS program document and asking personnel to read it and comment seldom rises to the top of anyone's priority list. The age old suggestion box, be it digital or hard copy is easily ignored. And tasks groups and focus groups are, well, task groups and focus groups.

This is a topic that is no different than many others where the input of the operations personnel is desired. The failure of so many methods is understandable once you put yourselves in the world of the operations teams. It is a pretty good bet that the majority of operations teams are not overstaffed. Then add to their normal role of maintaining and operating the facility all the other pulls on their time such as training, safety meetings, performance reviews, etc. Most offshore personnel have full schedules without being asked to review and comment on the SEMS program.

7.18.2 What good looks like

While I said that I had not seen what I would call a best practice, I have seen various pieces of what I think the core of a good EPP might look like. Those organizations who take the time to make sure that the operating personnel truly understand what SEMS and how it impacts specific roles and individuals have a sound basis for future discussions regarding improvements. The key here is to clarify that SEMS is not a separate program or initiative but is simply the work processes which direct the daily operations. This takes time and resources to make it clear that Element 4 is simply how the organization will handle managing change and is not an additional requirement. Understand that not every person needs to understand every element of the SEMS program. I have seen an operations leader spend significant personal time developing a matrix that she used to determine which roles needed to understand which elements. The organization then provided focused, efficient awareness training regarding SEMS.

With respect to getting input from operations personnel, I have observed organizations who did not attempt to make this an "event". While it is true that operators are required to annually review their SEMS program, I have yet to see an annual event that was successful in generating much feedback from the operations team. What I have seen are operators who use the normal work processes and meetings as opportunities to solicit input for improvements. This was evident when reviewing the SEMS program document and there were multiple edits not tied to annual program reviews. The edits were linked to improvements and changes pointed out by the operations personnel as they implemented the processes within the SEMS program. This ongoing input process takes a lot of effort by those responsible for the implementation of the SEMS program, but appears to be the best way to get input.

7.19 Element seventeen — reporting unsafe working conditions

§ 250.1933 What procedures must be included for reporting unsafe working conditions?
(a) Your SEMS program must include procedures for all personnel to report unsafe working conditions in accordance with § 250.193. These procedures must take into account applicable USCG reporting requirements for unsafe working conditions.
(b) You must post a notice at the place of employment in a visible location frequently visited by personnel that contains the reporting information in

7.19.1 Understanding element seventeen

This should be another free space on the Bingo card element. This element is satisfied by prominently posting the contact information where personnel can report unsafe working conditions to BSEE. The intent of this element is to assure that personnel who, for whatever reason, feel that they either cannot report unsafe conditions to the onsite supervision or who feel the onsite supervision is not addressing the issue. So, get the signs, post the signs, and talk about the signs in the facility orientation.

BSEE has recently made this "fool proof" via their Safety Alert No. 357, dated July 1, 2019. Within the subject alert is a sample sign that fulfills the requirements of 30 CFR 250.1933. I strongly recommend that you go to the BSEE website, find Safety Alert No. 357 and make sure your signage is in compliance. I do not include a copy here because you need to go to the BSEE website and review the most recent requirements and example.

7.19.2 What good looks like

The sign or posting of the BSEE contact information needs to be exclusive to reporting unsafe working conditions. As appropriate I have seen the Coast Guard contact information sharing the space but that is as far as I would take this. This is important enough to warrant its own posting. I have witnessed operators including this information on the same posting or sign with the UWA, SWA reminders, etc. This does not get favorable comments from BSEE representatives, and I understand why.

Suggested reading

30 CFR Part 254 — Oil Spill Response Requirements for Facilities Located Seaward of the Coast Line

Bureau of Safety and Environmental Enforcement (BSEE) website, https://www.bsee.gov/.

Center for Offshore Safety Website, https://www.centerforoffshoresafety.org/.

Code of Federal Regulations, 30 CFR 250, Oil and Gas and Sulfur Operations in the Outer Continental Shelf, Subpart A, Subgroup 76 Section 250.193, Reports and Investigations of Possible Violations.

Code of Federal Regulations, 30 CFR 250, Oil and Gas and Sulfur Operations in the Outer Continental Shelf, Subpart S, Safety and Environmental Management Systems (SEMS) (7-1-13 edition).

Recommended Practices for Development of a Safety and Environmental Management Program for OCS Operations and Facilities, American Petroleum Institute Recommended Practice 75, May 2008 edition.

CHAPTER 8

The SEMS audit

Contents

Shutterstock #1165591879.

The purpose of implementing a SEMS Program is not simply to satisfy BSEE and to pass the required audits. It is to reduce risk. But, the reality is that BSEE requires every OCS operator to conduct a third party audit. By "third party audit" I mean an audit that is conducted by someone other than the operator. Details of who this third party is

An Operations Guide to Safety and Environmental Management Systems (SEMS)
DOI: https://doi.org/10.1016/B978-0-12-820040-7.00008-9

required to be are coming later in this chapter. The key here is that the audit is mandatory, and the operator pays for the audit. What follows in this chapter is a discussion regarding how to get the most from your expenditure of resources. Not only does the operator pay for the audit, the process requires the dedication and utilization of significant operator resources. While audits would never be considered fun, they do not have to be dreaded or avoided; you do not need to find out the audit dates in order to concurrently schedule your vacation! Conversely, going into the audit unprepared or confrontational only serves to waste resources and dilutes any benefits of the audit. Accept the inevitability of the SEMS audit and follow some basic principles to get the most from the resources expended.

8.1 The BSEE ground rules

First off, know and understand the audit process required by BSEE. The audit requirements are included in 30 CFR 250 Subpart S, § 250.1920, as Element 12 and discussed in Chapter 7, Section 7.14 of this book. While not a substitute for a thorough understanding of the SEMS audit requirements, some key items that even those on the periphery of the audit need to understand are:

- The audit must be conducted by an accredited service provider (ASP)
- The Center for Offshore Safety (COS) is the accreditation organization and the list of ASPs can be found on the COS website (it does change)
- The audit will follow the COS protocol (COS-1-01)
- It is permissible to have personnel from the operators staff on the audit team but the lead auditor must be an employee, agent or representative of the ASP
- BSEE must approve your audit plan prior to initiation of the audit
- The audit frequency will be 3 years
- Each audit should encompass 15% of the facilities operated not to be repeated until all of the operators facilities are audited
- The audit report and the associated corrective action plan (CAP) will be submitted to BSEE

For those personnel who are integral in the planning of and the execution of the audit it is imperative that they fully understand 30 CFR 250 Subpart S, § 250.1920, API RP 75, and COS-1-01 as a minimum. Do not rely upon your audit provider to assure the process is followed; it is the operator's responsibility.

Second, know the deadlines. Operators have wasted time and resources because they did not plan their audit around the BSEE deadlines. The critical deadlines:

- Submittal of your written audit plan at least 30 days prior to the audit. In reality, get it done sooner than that. BSEE can, and does request modifications to audit plans and if you have made all the arrangements to have personnel available, SMEs on site, and facilities reserved it can be complicated and time consuming to change all of it.

- Submittal of the written audit report by the ASP within 60 days of the completion of the audit. While 60 days sounds like a lot of time, audit report preparation is complex and detailed. If you would like to be involved in the preparation of the report as well as to have the opportunity to see the report before it is submitted to BSEE, make your resources available to the ASP during these 60 days.

- Submittal of your Corrective Action Plan (CAP) within 60 days of the audit completion date. This is the responsibility of the operator, not the ASP. Many times this is the most difficult of the deadlines to meet, as it requires the operator to determine how best to address the deficiencies identified in the audit, what resources will be required, who will be responsible for each action item and how long it will take to complete. This is then submitted to BSEE for review. In developing the CAP, remember not all deficiencies are of equivalent risk, so prioritize appropriately.

- Audit frequency (unless otherwise requested by BSEE) is currently every 3 years. It does not look good if you have 3 years to be ready for the next audit and you are late in completing it!

8.2 Participate in the planning

It is tempting to simply ask the ASP to develop the audit plan independently, and bring it to you for review. This usually results in a "generic" but acceptable audit plan, but rarely results in optimizing the benefit the operator gets from the audit. Your goal should not be to "get through" the audit, but to maximize the useful information you get from the audit. This requires a mutual planning effort with the ASP which considers:

- Known gaps in your SEMS program. Why spend audit time and resources discussing gaps you are aware of and are working to address?
- Audit team make up. The audit team members provided by the ASP likely have a variety of backgrounds, some more suited to auditing your specific operations than others. Be a smart consumer and request audit team members whose expertize best correlates to the operations being audited. This may have a small impact on the overall cost of the audit but will pay off in the results.
- Include sufficient time in the audit plan. It is very tempting to try to get the audit over with as quickly as possible. This can result in a final report with more deficiencies than your operations may actually have. How does this happen? Suppose that during an audit the audit team discovers what appear to be deficiencies, but for which the operator actually has additional documents or information that demonstrates that these are not deficiencies. The audit schedule needs to allow your audit participants time to gather follow up information and present it to the audit team. Otherwise, the audit team has no choice but to include these as deficiencies in the report. I have experienced this in essentially every audit I have been involved with and as an auditor I would prefer to have time to review additional information to including an adverse finding in the audit report which is, in fact, not a finding.

8.3 Prepare for the audit

As an audit team member I have witnessed significant wasted resources and time due to the operator's audit participants being unprepared. Have all the documents you may be required to provide and discuss easily accessible. And remember, some documents may require legal review and editing prior to allowing the audit team to review them. Similar to having the documents available, have all personnel who may be required available as well. Delays in audit progress while the operator locates the SME for a particular topic is a preventable waste of time.

Prepare the audit participants with respect to what to expect during the audit interviews. Preparation does not mean a 30 minute information session the day prior to the audit. For those individuals who will be the primary participants in the audit, this can mean days of preparation. They need to understand the audit schedule, who will be in the room, and what all participants individual roles will be. Have a designated spokesperson for each topic and discuss this with all participants. Having the operator's participants all trying to speak at once or looking silently at each other to see who will answer creates an awkward and inefficient atmosphere in the audit room. Audit participants need to be familiar with the materials that may be reviewed during the audit and how to access them. It is a good idea to have the participants for each element review the COS audit protocol and be familiar with how the operator has implemented the SEMS program to satisfy the requirements of each protocol item. This does not mean trying to script a response or trying to avoid specific items or issues. In fact, when preparing the audit participants it is important that everyone is aware of any gaps that are known to exist and any actions underway to address them. As a participant in an audit on the operator's side of the table I have told an ASP's audit team at the start of an element session that the documents we would be showing were indeed due for review and that the review process was underway but not yet completed.

And sometime in every audit, the audit team will find what they think is a deficiency. Have your audit personnel prepared to hear this, and to react constructively. I have experienced firsthand the feeling when an auditor questions something that is in your area of responsibility. It is hard not to go into defensive mode, especially if there is pride of authorship. However, it is better to set the tone in the room as one of listening, understanding and discussing. Then provide the additional information to demonstrate compliance or accept there is room for improvement and move on.

8.4 Understanding the auditors

Take some time in the beginning of the audit to review and facilitate an understanding of the audit team members' backgrounds as well as how they view their purpose and goals for the audit. Some auditors may have little or no experience in offshore operations. This does not necessarily mean they cannot perform as part of the audit team, but it may require effort on the part of the operator to assure effective communications are occurring between the audit team and the operator's personnel. Auditors tend to speak in auditor language which does not always translate well into operations language. Having some participants in the room who have both audit and operations experience can go a long way to identifying when a discussion is being impacted by communications issues.

SEMS audit teams live and work in a world where there are currently 17 elements. Some operators have management systems that have more or less than 17 elements. Do not expect the audit team to learn and understand your management system. Map your management system elements to the 17 SEMS elements to avoid wasting time during the audit trying to teach the audit team the structure of your specific management system.

I have witnessed from both sides of the audit table confusion among the operator's participants regarding why the auditors are asking the questions they are asking. Even the best prepared audit participants can leave an interview session having discussed in detail an aspect of a specific element that they did not think warranted that much attention and furthermore having not been asked about what they thought was the most important aspect. The audit team has a limited time to get a "snapshot" evaluation of the operator's practices and procedures. They will focus their interviews on areas and topics where they may suspect deficiencies, not in the hopes of making the operator look bad but to develop an impression of the quality of the SEMS implementation. They will not likely ask a lot about those topics where the operator is obviously in compliance. This can be a bit of a letdown when the operator is anticipating a chance to demonstrate a particular aspect that they do very well. Most auditors think in terms of three primary questions that will tend to dictate the direction of the discussion:

- Does the operator have documented procedures addressing the SEMS element being evaluated?
- Do the documented procedures meet the SEMS requirements?
- Can the operator demonstrate that the documented procedures are being followed?

The rude, combative auditor does exist and I have experienced this on both sides of the audit table. This does not mean you have to accept the behavior and endure. What you do need to do is document the issues, and discuss them with either the ASPs audit team leader or the ASP's management. Very seldom does a full on confrontation in the audit room do anything but make the situation more unbearable. Not every individual is right for every audit.

8.5 Conducting the audit

Shutterstock #659261053.

When it comes time to actually start the audit, it is no time to relax and let the process go untended. I have already discussed having sufficient time in the schedule for obtaining and presenting additional information. If possible have some extra rooms where breakout discussions can take place and not take away from the progress of the audit. If the audit includes a visit to the facility (which it should), logistics are important. It is not easy to get an audit team and the operator's audit participants to an offshore facility and house them for multiple days. Helicopters need to be scheduled and bed space made available. Space to conduct interviews is at a premium offshore, and likely disrupts the usual meeting cadence. Have the space and meeting arrangements worked out prior to going offshore to optimize the limited time available.

If you remember nothing from this section, remember this; never send an audit team offshore without providing qualified escorts or confirming the availability of the offshore personnel to accompany the auditors. It is tempting to save some bed space by having the offshore staff that is on the current hitch conduct the field portion independently, but the results are often not good, specifically on large facilities. This comes back to the part about understanding the auditors. Having effective auditor to operator communications is just as hard offshore as in the office so your "translators" need to be there as well. Operations staff trying to juggle their normal roles as well as participate in the audit can become stretched thin. Further, some audit team members may understand very little about the organizational structure offshore and ask someone a question whose role does not require them to have any knowledge of that topic. This may appear to the audit team as a deficiency, but they actually just asked the wrong person. Finally, the escorts provided by the operator need to be accountable for the safety of the audit team. Do not allow the audit team to wander around the facility independently, expecting the busy offshore staff to watch out for them.

Suggested reading

Center for Offshore Safety website, https://www.centerforoffshoresafety.org/.
Code of Federal Regulation, 30 CFR 250, Oil and Gas and Sulfur Operations in the Outer Continental Shelf, Subpart S, Safety and Environmental Management Systems (SEMS) (7-1-13 edition)
Recommended Practices for Development of a Safety and Environmental Management Program for OCS Operations and Facilities, American Petroleum Institute Recommended Practice 75, May 2008 edition.

CHAPTER 9

Contractors

Contents

Shutterstock #1017247639.

An Operations Guide to Safety and Environmental Management Systems (SEMS)
DOI: https://doi.org/10.1016/B978-0-12-820040-7.00009-0

9.1 What do the regulations say?

§ 250.1914 – What criteria must be documented in my SEMS program for safe work practices and contractor selection?

Your SEMS program must establish and implement safe work practices designed to minimize the risks associated with operations, maintenance, modification activities, and the handling of materials and substances that could affect safety or the environment. Your SEMS program must also document contractor selection criteria. When selecting a contractor, you must obtain and evaluate information regarding the contractor's safety record and environmental performance. You must ensure that contractors have their own written safe work practices. Contractors may adopt appropriate sections of your SEMS program. You and your contractor must document an agreement on appropriate contractor safety and environmental policies and practices before the contractor begins work at your facilities.

(a) A contractor is anyone performing work for you. However, these requirements do not apply to contractors providing domestic services to you or other contractors. Domestic services include janitorial work, food and beverage service, laundry service, housekeeping, and similar activities.

(b) You must document that your contracted employees are knowledgeable and experienced in the work practices necessary to perform their job in a safe and environmentally sound manner.

Documentation of each contracted employee's expertize to perform his/her job and a copy of the contractor's safety policies and procedures must be made available to the operator and BSEE upon request.

(c) Your SEMS program must include procedures and verification for selecting a contractor as follows:

(1) Your SEMS program must have procedures that verify that contractors are conducting their activities in accordance with your SEMS program.

(2) You are responsible for making certain that contractors have the skills and knowledge to perform their assigned duties and are conducting these activities in accordance with the requirements in your SEMS program.

(3) You must make the results of your verification for selecting contractors available to BSEE upon request.

(d) Your SEMS program must include procedures and verification that contractor personnel understand and can perform their assigned duties for activities for activities such as, but not limited to:

(1) Installation, maintenance, or repair of equipment;

(2) Construction, startup, and operation of your facilities;

(3) Turnaround operations;

(4) *Major renovation; or*

(5) *Specialty work.*

(e) *You must:*

(1) *Perform periodic evaluations of the performance of contract employees that verifies they are fulfilling their obligations, and*

(2) *Maintain a contractor employee injury and illness log for 2 years related to the contractor's work in the operation area, and include this information on Form BSEE—0131.*

(f) *You must inform your contractors of any known hazards at the facility they are working on including, but not limited to fires, explosions, slips, trips, falls, other injuries, and hazards associated with lifting operations.*

(g) *You must develop and implement safe work practices to control the presence, entrance, and exit of contract employees in operation areas.*

As a contractor working in the OCS, there are two distinct points of view you can take with respect to your SEMS role and obligations. You could read 30 CFR 250 Subpart S § 250.1914 and decide that SEMS really does not apply to you. The accountability for making sure that the people working at any task on the facility are competent and are working within the Operator's SEMS program lands pretty squarely on the facility operator. So you could decide not to concern yourself with SEMS and let the facility operators handle it.

At first look, this approach may seem the way to go, after all it does not require the contractor to learn about SEMS, develop any new processes or procedures or hire any specialists; all of which cost money and impact the bottom line. And the operator's are always pushing contractors for cost control. Here is where following what at first seems to be the optimum approach can lead to unintended consequences that may impact a contractor's ability to compete for work. To understand why, contractors need to look at the situation from the operators' point of view. The operators are very aware that the accountability lies with them. It is re-enforced every visit from BSEE and every audit. I need to be clear; I am not making any judgment regarding how the regulations read on this topic; that is a discussion that could keep industry experts busy for days. For now, just take it as it is and effectively work within it.

The operator has a complex facility operating within a limited amount of space offshore that can have 200 + people on board at any time. There are hydrocarbons present, there are hazardous materials, there may be multiple operations and projects ongoing simultaneously, and to top it off a helicopter periodically lands and takes off from there. Piece of cake, right? As the operator is evaluating contractors to participate in this complex world, they will likely inform the contractor that activities on the facility must be done in accordance with the operator's SEMS program. If this discussion elicits a blank look from the contractor's personnel, it is easy for the operator to conclude that this particular contractor will need significant oversight by the operators' personnel to assure that work is done per the SEMS program. This can be seen by the operator as a net increase to the operator's required staff level or a diversion of limited resources. Conversely, if a contractor is able to demonstrate an understanding of SEMS, and discuss the SEMS criteria that they will be required to work within there is a potential competitive advantage.

For the doubters reading this, here is another thought to ponder. Let's say a contractor whose SEMS knowledge is minimal successfully gets work on a facility covered by SEMS regulations. Then say an incident involving this contractor's personnel occurs and the investigation leads to the conclusion that the SEMS program was not being followed. This is not good for anybody involved. While the contractor may fall back on the fact that the operator is accountable for adherence to the SEMS program, it is very likely that the contractor's personnel are participants in any reviews and included in any documentation from the incident. Like it or not, as a contractor you are involved.

The purpose of this book is not to instruct you how to successfully pass an audit but rather how to use the SEMS requirements to reduce risk. However, the audits are a part of operating in the OCS. During the required third party SEMS audits, it is the operator who is being audited. As a contractor, do not let this result in your not taking the audits and the audit process very seriously. As an audit team member I have interviewed a lot of contractor personnel as in many cases key personnel with knowledge of how operations are managed on a facility are contractor staff. It can cause quite a panic among the staff when the audit team asks for a contractor's employee to come to the conference room to be interviewed. Not to mention the fact that now the audit team is getting information from what may be a nervous and unprepared person. I have also been on audit teams where the contractor's role was significant enough it warranted spending time conducting interviews at the contractor's office and on the contractor's facility. Make no mistake, just because the contractor may not be mentioned by name or that the audit report and the associated corrective action plan belong to the operator, contractors can and are involved in the audits and can become indirectly associated with audit findings.

9.2 What should a contractor do?

Shutterstock #736797094.

First and foremost contractors working in the OCS need to educate their staff regarding SEMS. I have heard from some small contractors that they just don't have the resources for this. The answer is to educate "fit for purpose". The breadth and depth of SEMS knowledge needed is very different for a contractor supplying a few personnel for a compressor repair than it is for a contractor supplying a drill ship or a process skid.

A good first pass at determining the extent of the education required is to simply go through the 17 Elements of SEMS and see how many actually impact the contractor's personnel. Remember, the 17 Elements are intended to cover a comprehensive safety and environmental management system for the operator. Contractors may not have a role to play in all 17 Elements. Focus the education on those elements that impact the work and services provided by the contractor.

Contractors are playing "catch up" when it comes to SEMS education in that many operators have spent significant time educating their own personnel, sometimes without any attention paid to contractors. This is both good and bad. Obviously, not understanding SEMS is the bad part. The good part is there are competent subject matter experts (SME) available to contractors in the form of specialists who have developed a thorough understanding of SEMS since the inception in 2010 and can help identify what the contractor staff needs to know as well as tailor training to fit the contractor's needs. Sometimes it is not bad to arrive at the party a little late, and in this case it means there are resources available to the contractors with experience developed over the last almost nine years. It is obviously the decision of the individual contractor, but there may be a savings in time and resources to warrant utilization of an SME.

9.3 Contractor SEMS programs

Shutterstock #772014820.

Major contractors doing significant business in the OCS should consider developing and implementing their own SEMS program. I want to emphasize the implementation part. Having a SEMS program is only an advantage if the contractor can demonstrate that their personnel understand and follow the program. This may seem like a significant dedication of resources for something the operator is ultimately accountable for, but once again hold off on that initial reaction. It is time to see this from the operator's point of view once again.

When a major contractor comes to the table with their own SEMS program, the process for completing the contracts and getting on with the work is simplified. And to the operator time is money, sometimes a lot of money. As discussed in detail in Chapter 7, Section 7.8, when the contractor has a

SEMS program, a Bridging Document can be developed that "bridges" between the operator's SEMS program and the contractor's SEMS program. This simply means that the two organizations sit down and evaluate the two programs and choose which processes are preferred for the work being contracted for. All personnel, both contractor and operator, are then instructed on the specifics of each element per the bridging document.

Developing a SEMS program for a contractor may be slightly less complex than developing a SEMS program for the operator. The contractor SEMS program does not require submittal to BSEE and not all elements may apply to each contractor. Once again, being late to the party can be good, as there are SME resources to guide contractors through the process of developing and implementing a SEMS program, and to assure such a program is limited to those elements relevant to the specific contractor.

9.4 The elements — a contractor's view

Shutterstock #551512852.

I want to be very clear who this section is and is not written for. It is not written for the large contractor who determines that their best approach is to develop their own SEMS program as described in Section 9.3. This section is for those contractors who are not large enough or of a scope of business that it warrants development of a full SEMS program, but yet need to understand SEMS to be a competitive and effective contractor in the OCS.

Looking at the individual 17 SEMS Elements from a contractor point of view is a bit tricky, in that it varies significantly by the type and scope of the work the contractor does and what resources the contractor provides. For major contractors, such as drilling contractors, it is pretty straightforward. If a contractor reviews the 17 elements and most of them apply, it follows that approaching SEMS similarly to an operator makes sense. This requires a full understanding of the requirements and likely benefits from the development of a fully documented SEMS program. Examples are drilling contractors or contractors providing operating staff. For these contractors, Chapter 7, "The SEMS Elements" is an applicable resource.

That leaves a lot of other contractors in the uncomfortable situation of not really needing a full SEMS program, but needing to understand and work within a variety of operators' SEMS programs. It is for these contractors that the following element specific sections are written. These are by no means the only things a contractor may need to understand or be aware of, but it should provide a solid foundation upon which a contractor can base discussions and planning with the operators. I have omitted elements 1 and 16 in this discussion as they are directed primarily at the operator. Element 1, General, and Element 16, Employee Participation Program would be of interest to those contractors who develop their own SEMS program and the requirements would parallel those for the operator's SEMS Program (Chapter 7, The SEMS Elements).

9.4.1 Element two — safety and environmental information

If a contractor brings equipment onto an operator's facility, there is a component of Element 2 that needs to be considered. SEMS Environmental and Safety Information (S&EI) is the information required for hazard analysis, operating procedures, emergency response and incident investigation, among other things. Consequently, a contractor bringing equipment onto the operator's facility needs to have the appropriate information for that equipment. This may be flow diagrams, operating procedures, operating limits, etc. The addition of equipment, even if temporary, may well trigger the operator's management of change process (MOC), in which case the operator will need more information such as equipment weight, emissions, etc

9.4.2 Element three – hazard analysis

Contractors need to understand the concept of a JSA. By understand I don't mean that the contractor personnel understand that they have to sign a JSA before they start working. They need to understand "why" they are going through the process and give it the attention it warrants. There is an all too familiar scene where the operator's representative is going through the JSA while the contractor's personnel are already getting their tools ready, checking materials, but not listening. Then they all line up, sign the JSA form and go to work. I have reviewed incidents where not following the JSA contributed to an incident occurring shortly after the review of the JSA. I have observed JSAs that had little detail or substance, obviously done quickly to satisfy the requirement of completing a JSA. I am aware of workers signing off on a JSA that was written in and discussed in a language other than their primary language.

Contractors should be aware that some of the JSA process is specified in the SEMS requirements. The operator should require that the immediate supervisor of the crew performing the job onsite conducts the JSA, signs the JSA, and takes steps to ensure that all personnel participating in the job understand and sign the JSA. This is not a reflection of the contractor, the crew or their qualifications; it is required by SEMS. As a contractor if an operator does not follow these steps it should be a concern.

The JSA is an opportunity for contractor personnel who may not be familiar with the facility to review hazards with operator personnel who know the facility in detail. Contractor personnel need to see the JSA process as a significant tool they have to reduce the risk associated with the tasks they are contracted to do. The JSA process needs to be taken seriously, to the extent that if the operator does not perform a thorough JSA the contractor personnel recognize it and do not proceed until an acceptable JSA is completed. Take responsibility for your own safety and risk mitigation.

9.4.3 Element four – management of change

This element carries personal significance in this context for me, and provides me with an example that illustrates the need for contractors to understand the concept and use of a management of change (MOC) process. I led an investigation into an incident that had injured several contactor employees. It was a facility hydrotest; something the contractor's personnel on site had done many times. As the piping for the test was being hooked up, the contractor's personnel noticed that the fitting on the pipe they were connecting to did not look like what was depicted on the drawings they had been provided. The contractor personnel went forward with the procedure, connected to the pipe, and began to increase pressure according to the hydrotest procedure. Contractor personnel subsequently observed a small leak where their equipment was connected to the pipe, and left the control trailer to attempt to see what was leaking. It was at this point that the fitting where the test equipment hooked to the pipe failed, sending people and equipment in multiple directions.

The investigation revealed that the fitting that had been noticed by those on site as being different than what was shown on their drawings had been fabricated and installed during construction to help in the process of getting the line into position, and had been left in place. What has this got to do with MOC understanding? The investigation, as is common with most incidents, concluded multiple factors contributed to the incident, but among them was the fact that no one questioned the lack of an MOC associated with the unexpected fitting. Contractor personnel need to understand why the MOC process is utilized and when not to proceed without the appropriate MOC. Do not simply trust that the operator or another contractor will implement the MOC process if it is needed, and don't be concerned about questioning the operator, even if it means a delay in the work. Be well versed in the use of the MOC process and know when to question a possible lack of MOC implementation. In the example, if one person on the location had requested to see an MOC for the new fitting it is possible the incident would have been avoided.

9.4.4 Element five — operating procedures

Obviously, if a contractor is bringing equipment onto the operator's facility, the operating procedures for the equipment need to be provided as well. There is another aspect of operating procedures that a contractor needs to be aware of. SEMS requires that the operator have operating procedures for a variety of activities, including:

- Start up
- Shut down
- By passing out of service equipment
- Emergency operations
- Prevention of exposure to and handling of chemicals

This means that if the operator requests contractor personnel to partic-ipate in shutting down or isolating equipment prior to work starting or to participate in starting up equipment after the work is completed, there should be a procedure to follow. As a contractor, insist on this regardless of how simple or routine the task appears to be. For example, shutting down or starting up a pump may appear to be as simple as pushing a but-ton. However, offshore facilities are complex and what appears to be a simple task can result in significant impacts on other parts of the process. And offshore, the other parts of the process may be very close to where you are working. It is very common for multiple activities to be ongoing at the same time, requiring the operator to follow work permit and simul-taneous operations (SIMOPS) processes. The simple start up or shut down of the example pump could have a negative impact on other work crews. Bottom line, contractor personnel need to be aware that procedures are required for just about everything done on the facility and should insist on having such a procedure before taking any action.

9.4.5 Element six — safe work practices

As a contractor be aware that the operator must have some key safe work practices (SWP) and a permit process in place. The key here is that they are not called out specifically in 30 CFR 250 Subpart S, but are a SEMS requirement as 30 CFR 250 Subpart S includes by reference the require-ments of API Recommended Practice 75. These are critical and a con-tractor should understand if the services they provide are covered by these requirements and ensure that no covered tasks are performed without the implementation of these SWPs and the permit process.

At a minimum there should be documented SWPs that details effec-tive and appropriate methods for:
- Opening pressurized or energized equipment or piping;
- Lockout and tagout of electrical and mechanical energy sources;
- Hot work and other work involving ignition sources;
- Confined space entry;
- Crane operations;

Along with these SWPs there must be an effective permit to work sys-tem. At a minimum a permit to work should be issued before work involving the following areas is initiated:
- Opening pressurized or energized equipment or piping;
- Lockout and tagout of electrical and mechanical energy sources;

- Hot work and other work involving ignition sources;
- Confined space entry;

This is an element where contractor education can go a long way to assuring incident free working conditions. Contractor personnel should be keenly aware when what they are doing is associated with these SWPs and requires effective implementation of a permit process. If a situation arises where proper methods do not appear to be followed, the work should be halted until all parties are comfortable with the arrangements. Contractor personnel should be sufficiently trained to recognize a potentially dangerous situation and to confidently discuss this with the operator.

Operators may have additional SWPs that the contractor is expected to understand and comply with. Make sure all affected personnel understand these SWPs and have an opportunity to discuss and ask questions with the operator's representative before the work starts.

9.4.6 Element six – contractor management

Here is another one of those places where it would be easy to place all the responsibility on the operator to make sure that the contractor is working according to the operators SEMS program. Indeed, that is pretty much how the requirements are written. It can be resource intensive for an operator to accomplish this when there are multiple contractors doing various jobs on a facility. As an audit team member I have observed thorough operators with fully detailed SEMS programs struggle to keep up with this task. So, why should a contractor care? The SEMS program is designed to minimize risk. Therefore, if a contractor is working according to the SEMS program, their workers should be exposed to a lower risk environment. This is everyone's goal, operator and contractor alike.

As a contractor if you observe that there are no operator representatives or processes in place to assure that your personnel are working within the SEMS requirements, you need to provide assurance that your personnel are working within the SEMS requirements. This may entail some discussion with the operator or the contractor assigning additional resources to provide the assurance. While this is an increased cost to the contractor, working outside of the SEMS requirements can increase risk which can lead to incidents. Not only is it smart from an incident reduction perspective, it just might get you on the preferred vendor list.

9.4.7 Element seven — training

The operator is responsible for assuring that the personnel on the facility are trained and competent for the work they are hired to do. Many times, this entails the operator requesting specific skill sets from the contractor who in turn agrees to provide such personnel. The contractor will likely supply the operator with the names and training credentials of the personnel that will be coming to the facility. Then comes the day when the contractor personnel arrive offshore and the operator discovers that the personnel who have arrived are not the same ones promised by the contractor and do not have the required competencies. The impact of this is significant, and as a contractor it is a sure way to get removed from the preferred vendors list. Think about it. The operator has just wasted seats on the helicopter coming to the platform and will have to pay for contractor personnel to return to shore as well as pay for flights to get the correct personnel to the facility. Further, if the competent personnel are not readily available, the work on the facility can be delayed, downtime extended, and revenue impacted. It is the contractor's responsibility to provide the competent personnel as promised, and while people get sick, take vacations, etc. a SEMS complaint operator will not initiate the work without proven competent contractor staff. In the event the competent personnel become unavailable, do not substitute personnel without consultation with the operator. From an operator perspective it is better to delay the work than to send the unqualified personnel offshore.

9.4.8 Element eight — mechanical integrity

This is very simple, yet very hard. The simple part is that if you as a contractor are providing equipment meeting the SEMS definition of critical equipment that will be placed on the offshore facility, regardless of the duration, you should provide the operator with full details of the mechanical integrity program and history for that equipment. The operator is responsible for the mechanical integrity of all critical equipment on the facility. According to SEMS, critical equipment is all equipment and systems used to prevent or mitigate uncontrolled releases of hydrocarbons, toxic substances or other materials that may cause environmental or safety consequences. If the failure of the equipment provided by the contractor meets this definition then the full scope of Element 8 (Chapter 7, Section 7.10) applies to that piece of equipment.

Now for the hard part. If a contractor supplies critical equipment, it is best for the contractor to take the time to fully understand Element 8. As a start, I would go to Chapter 7, Section 7.10, Element 8 — Mechanical Integrity and read and understand it. Similar to the recommendations in Section 7.10, the next step would be to enlist the services of an appropriate SME to develop and implement a mechanical integrity program specific to your equipment.

While this sounds like a lot of work and possibly expense, keep in mind that an operator should not allow a contractor to provide critical equipment that is not part of a comprehensive mechanical integrity program. In fact, the SEMS audit protocol (COS-1-01) has a question specific to this which the operator will be asked to answer. Also keep in mind that if the failure of a contractor's equipment results in a safety or environmental incident, the contractor will be involved in the incident investigation and all the documentation and reports generated by the investigation.

9.4.9 Element nine — pre start-up review

Many times the contractor's personnel are involved with the start up of repaired or new equipment and processes. As with safe work practices the pre start-up review (PSR) process should be transparent and fully involve contractor personnel who will be participating. Better yet, the contractor should be involved in the PSR process from start to finish. As a contractor, do not start up anything without fully understanding the PSR process that was completed. If the operator's representative gives a verbal approval to start anything, regardless of the scope or complexity, the contractor should ask to see and understand the PSR that was completed. Many times the contractor personnel have more experience than the operator in starting up specific equipment, and the contractor input is beneficial to the PSR. On a complicated facility such as an offshore platform, starting up anything has the potential for significant risk exposure if the preparation is not thorough.

If an operator did not complete a PSR or will not provide the PSR to the contractor for review, the contractor is potentially putting their personnel at risk by participating in the startup.

9.4.10 Element ten — emergency response and control

Know what to do in case of an emergency. It is that simple. When the operator gives the contractor staff the orientation which should include emergency procedures pay attention. And if the orientation does not include this, ask for it. It may be true that some of the personnel have heard similar orientations multiple times on multiple facilities, and it may be true that incidents are rare, but they do occur. And if you are not familiar with the platform you are on do not count on your intuition or the intervention by others to get you to the right place.

I can attest from personal experience that when the alarm is going off and waking you from a deep sleep you better have paid attention to your lifeboat assignment, the check in procedure and where your lifejacket is. Getting 200 + people to the correct lifeboat muster point, checked in and ready to evacuate is complex and seemed like controlled chaos to my recently awakened self. Further, not knowing what to do can put others at risk that must spend time locating you and getting you to the right place with the right equipment. I will admit that while I was able to navigate to my assigned lifeboat, and correctly check myself in, I had forgotten my life jacket which was still on the top of the closet I my room. Fortunately, one of the largest human beings I have ever seen slapped a life jacket on me with enough force that I will never make that mistake again. While I did not have to evacuate, getting to the correct lifeboat muster point, getting checked in and being in the right place to get off the facility should it have been necessary felt like an important accomplishment. After that experience, I not only pay attention in the orientations, I make sure I know the layout of the facility, my path to my muster area and where the lifejackets are.

9.4.11 Element eleven — investigation of incidents

When contract personnel are involved in an incident, generally the facility operator's incident investigation process is followed. This does not mean that contractor personnel are not involved. This can be a difficult position for contractor staff as incident interviews can feel intimidating and contractor staff may fear that their answers can put the contractor's reputation at risk. Having led multiple incident investigations I offer some suggestions to contractors.

First, it is important that contractor leadership take the initiative to become involved early in the process with the operator's investigation leader. The contractor leadership can then inform their personnel regarding how the investigation will proceed and what to expect. It is also good to confirm with the operator's leadership that the goal of the investigation is to prevent recurrence, not to place blame. It is important for both the contractor and operator personnel to feel confident about this. As soon as there is a hint of blame finding those involved become defensive and it becomes difficult to reconstruct what actually happened. Should you find yourself involved in an investigation where the operator does appear to be looking to place blame, my personal opinion is that this is not an operator whom I would want a long term relationship with.

Second, contractor leadership needs to be sure to stress to their staff to be fully transparent in their interviews and answers. The success of the investigation with respect to identifying the true root causes and the action plans to effectively mitigate them are dependent upon the information provided by those involved.

Finally, work as a partner with the operator in developing the actions to prevent a recurrence. This helps build a long term relationship between the operator and the contractor and send a message to workers on both staffs that this is important.

9.4.12 Element twelve − auditing

It is important for contractors to understand that while it is the operator who is the subject of the required SEMS audits, in many ways the contractor is a part of the audit, especially if the contractor is a significant supplier of personnel and/or services. As I discussed earlier in this section, contractor personnel can be included in the audit interviews, and to some extent in the audit findings. As a contractor, ask the operator to communicate to you any areas where there is potential for contractor personnel to be involved and ask for those personnel to be included in any audit preparations. I have interviewed contractor personnel during audits that were well versed n SEMS and their role, and others who had been called into the conference room to address a specific issue with no warning or preparation. While the audit is an audit of the operator, do not put your personnel in a position they are unprepared for and where a rushed response could be misunderstood.

9.4.13 Element thirteen − recordkeeping

For contractors this is short and simple. Don't dispose of any documents or paperwork after a job is finished, rather let the operator do this. Also, do not be offended when the operator asks to retain copies of documents. The operator has SEMS requirements for keeping specific documentation and may have additional internal requirements. When the job is over, hand what is hopefully a dirty, wrinkled packet of JSAs, procedures, permits, etc. to the appropriate operator's representative. As an operations manager, I viewed the dirty, wrinkled documents as a sign they were actually used at the job site and did not set on a desk in an office while the work was proceeding. If a contractor's policy is to keep these documents, then it is worth the effort to make copies such that the contractor and the operator both have copies.

9.4.14 Element fourteen — stop work authority

This is one of, if not the most important element for contractors. The operator is required to have a stop the work authority (SWA) program as part of their SEMS program and to take time to familiarize all contractor personnel with it. The operator is also required to include stop work authority wording on many documents the contractor will see, including the JSA. This should be a topic of the first orientation contractor personnel receive when arriving at the facility. If it is not, the contractor leadership should ask for a review of this before the work is initiated.

The contractor also needs to fully embrace the concept of stopping work when anyone involved has a concern. Delays cost money, and there is no getting around that. They cause schedule problems that are compounded in the offshore world where delays result in crews missing helicopter flights and subsequent jobs are impacted. However, incidents can cost so much more. They can cost pain and suffering as well as money. Do not let the pressure of schedules become an impediment to an effective stop the job culture. This can be difficult for contractors where personnel are working for various operators under various SWA programs. The best approach is to make sure all contractor personnel know they can stop the job regardless of the operator, what personnel are involved in the work, or the schedule pressures. My personal example is two contract employees opening packaging in a warehouse. The materials were needed on the work site for work to continue and the contractor's employees wanted to inspect them prior to loading them for transport to the site. Bottom line is one person used the wrong tool to open the packaging, cut themselves and required a trip to the emergency room. In the incident investigation the non injured employee indicated he thought about stopping the work to get a better tool but knew they were in a hurry. The result to me as an operator was a day delay in progress and a recordable injury.

9.4.15 Element fifteen — ultimate work authority

Another relatively easy Element for most contractors, this should be a topic in the initial orientation as well as posted in multiple locations on the facility. A good self check is at no time should any member of the contractor's staff have any question regarding who the ultimate work authority (UWA) is or where to quickly confirm this. So why should contractor personnel even care about this? The UWA has many specific roles including but not limited to approving returning to work after a stop the work event, response to incidents and emergency response.

9.4.16 Element seventeen — reporting unsafe working conditions

This is a "safety valve" for the assurance of safe work practices and environments, sometimes referred to as the "whistleblower" element. Should contractor personnel find themselves in the situation where they are being asked to work in what they feel is an unsafe manner, and the onsite supervision is unwilling to listen or respond there should be posted on the facility contact information for BSEE, and USCG. Contractor leadership needs to assure that their employees are aware of these postings and how to utilize the information on them. Contractor personnel should also understand that this is an anonymous process.

Suggested reading

Bureau of Safety and Environmental Enforcement (BSEE) website, https://www.bsee.gov/.
Center for Offshore Safety website, https://www.centerforoffshoresafety.org/.
Code of Federal Regulations, 30 CFR 250, Oil and Gas and Sulfur Operations in the Outer Continental Shelf, Subpart S, Safety and Environmental Management Systems (SEMS) (7-1-13 edition).
Recommended Practices for Development of a Safety and Environmental Management Program for OCS Operations and Facilities, American Petroleum Institute Recommended Practice 75, May 2008 edition.

CHAPTER 10

BSEE SEMS and the UK Safety Case

Contents

The purpose of this chapter is not to provide a full comparison between the BSEE SEMS program and the UK Safety Case program, nor is it to make any conclusion regarding the relative merits of either system. There are two purposes for this chapter. First, there are a significant number of people working in the US OCS who have little to no understanding of what the Safety Case is or where it came from. This chapter should at least give you the basics regarding the development and where to look if you need or want more details. Second, I have worked with many individuals who have come to the US OCS with significant experience in the Safety Case system and find themselves frustrated in trying to understand where this thing called SEMS came from and who the players are. This chapter will give you some basics as well as my observations regarding those people who have made the transition most efficiently.

Before going into this section, let me be fully transparent; I have never worked in the North Sea nor have I have never worked in an area that has adopted the North Sea Safety Case system. My level of understanding is built upon my reading of the regulations and countless discussions with people having experience in the North Sea working under the UK Regulations. I have had the opportunity to include in my resource group the unique and valuable perspectives of one of the 61 survivors of the Piper Alpha incident which occurred on July 8, 1988 and claimed the lives of 167 workers. It was the Piper Alpha incident which prompted the UK to initiate work regarding the improvement of safety in offshore operations.

The Cullen Inquiry was set up in November 1988, chaired by William Cullen, then a Scottish judge. The final report was released November 1990, and the recommendations of the report led to the

An Operations Guide to Safety and Environmental Management Systems (SEMS)
DOI: https://doi.org/10.1016/B978-0-12-820040-7.00010-7

development of The Offshore Installations (Safety Case) Regulations 1992. The Safety Case Regulations have been revised in 2005 and 2015, with each succeeding revision including more detail and clarifications. Consequently, there are offshore personnel who have worked within the Safety Case Regulations for over 20 years.

The BSEE SEMS regulations were enacted after the Deepwater Horizon incident in the Gulf of Mexico, which occurred in April 2010. The Deepwater Horizon incident claimed the lives of 11 people and caused the largest oil spill in history. The Workplace Safety Rule became effective November 15, 2010, which is referred to as SEMS I. The SEMS II revisions became effective June 4, 2013. For the details of the history of the BSEE SEMS development, please refer to the Chapter 2, The History of SEMS. What follows in this chapter is a brief summary of the development of SEMS, which is intended to provide a quick informational base for those new to the US OCS. At the most basic level, understand that BSEE SEMS regulations are based upon the American Petroleum Institute Recommended Practice 75 (API RP 75), "Recommended Practice for Development of a Safety and Environmental Management Program for Offshore Operations and Facilities". This is not to say that the UK Safety Case work and experience was not an influence, but the primary resource for the development of the BSEE SEMS was APIs RP 75.

Since the inception of the BSEE SEMS requirements in 2010, industry groups, HSE interest groups, workshops, etc. have discussed to varying levels of detail the difference between the UK and BSEE Regulations. Usually these discussions quickly turn to which is preferable, and then proceeds to why the US chose to independently develop a program when the UKs Safety Case regulations have been in place for an extended period of time. Somewhere in this part of the discussions the group moderator ends up taking charge and trying to get the discussion moving in a productive direction as there can be as many opinions on the above questions as there are people in the room. Indeed, I have developed my own opinion and will gladly share it over a beverage or a meal, but I am not including my opinion in this book. My opinion is no better or worse than anyone else's, and it does not contribute to the purpose of this section.

The offshore energy industry is a global industry, and the movement of individuals from one part of the world to another is common. As assets are developed, depleted, divested, etc. personnel with technical, operational and managerial experience are re-assigned to different assets and production areas. Consequently, we find operating staffs comprised of people who are experienced working within the Safety Case system and those experienced working within the SEMS system working together, often interspersed with personnel who are not familiar with either system. The resulting dialogue can be both confusing and frustrating. Who is "Bessie" and why does she have so much influence? Isn't the Safety Case the box on the wall with the first aid supplies in it? Do I need an RP75 or an L154, and what kind of a part is it?

Think about timing of the regulations before going further. The Offshore Installations (Safety Case) Regulations was enacted in 1992. The Workplace Safety Rule (SEMS I) was enacted in 2010. The Safety Case has had a significant head start over the BSEE SEMS program. In 2018 there are people who have 20 + years of experience working within the Safety Case requirements. In contrast, the BSEE SEMS requirements have been in place for less than ten years. I have witnessed people with significant experience working within the relatively mature Safety Case system finding themselves working within the relatively young BSEE SEMS system. To an outsider this may not seem like a big deal, but for those in offshore operations it can make the transition to that new role difficult. While the basic objectives and goals of any operation are generally similar, the operations are likely guided by an Operations Management System (OMS). The OMS details "how" work is done, the processes to be followed, the documentation requirements, etc. This is the tool utilized to produce consistency in operations and provide assurance of risk mitigation and compliance with regulations. I like to draw a parallel with auto racing (because I am a fan). Regardless of the category of racing (Formula 1, NASCAR, SCCA, etc.) the goal is to win races while maximizing driver safety. The drivers are very skilled at and knowledgeable regarding the mechanics and technique of driving race cars. However, when we see drivers move from one type of auto racing to another it is rare to see a smooth transition. Even though the basic goals are the same and the driver is skilled, the system within which the driver now operates is different. Similarly, if you have worked in a system where the OMS was tied to the Safety Case it will likely look and feel different than one that is tied to SEMS.

Once again, I need to be fully transparent. The rest of this chapter is wholly based upon my own experiences over the last ten years working as a consultant regarding the SEMS requirements. This has included multiple instances of assisting those new to the BSEE SEMS requirements in becoming comfortable working within them, as well as countless discussions regarding the differences between the Safety Case and BSEE SEMS requirements. The remainder of this chapter is focused on helping individuals who are making the transition from the Safety Case to the BSEE SEMS system.

The history and development of the Safety Case may be very familiar to you and the understanding of where it came from and how it evolved into the current form may be well known to you as well. There is a comfort in familiarity. As tedious as it may be, it is important to understand how and why the BSEE SEMS requirements came to be. It is easy to casually state that SEMS was developed because of the Deepwater Horizon incident and the Safety Case was developed because of the Piper Alpha incident. While these statements are true, the stories are more complex than that. Remember, the Safety Case came from the study prompted by Piper Alpha, but has gone through multiple revisions and refinements. SEMS, while a much newer system, has also gone through revisions and refinements.

The Bureau of Safety and Environmental Enforcement (BSEE) was not created immediately after the Deepwater Horizon incident, nor was it the first agency responsible for oversight of US offshore drilling and development. In the wake of the incident, the existing agency, the Minerals Management System (MMS), was subjected to significant scrutiny regarding the issuance of drilling permits and the effectiveness of the organization. The result was the May 19, 2010 replacement of the MMS with the Bureau of Ocean Energy Management, Regulation and Enforcement (BOEMRE). You may well find documents that refer to the BOEMRE or personnel who refer to "Boomer". The original Workplace Safety Rule (SEMS I) became effective November 15, 2010. Almost a year and a half later, in October 2011, the BOEMRE was dissolved and replaced by; the Office of Natural Resources Revenue (ONNR), the Bureau of Ocean Energy Management (BOEM), and the Bureau of Safety and Environmental Enforcement (BSEE). This organizational structure is still in place at the writing of this book. Do not be surprised if you find documentation for an operation that begins with reference to the MMS, then the BOEMRE and finally to BSEE. This book is concerned only with BSEE, referred to as "Bessie".

You will also need to understand the role of the Center for Offshore Safety (COS). People new to operating within the U.S. OCS many times assume COS is a division of BSEE. It is not. They may think it is a government agency. It is not. They may think it is a part of one of the various other worker safety organizations such as OSHA. It is not.

So what is it? The Center for Offshore Safety was chartered by the American Petroleum Institute (API) in 2011 in response to recommendations included in the Presidential Oil Spill Commission investigation into the Deepwater Horizon Incident. The API is a primary trade association for all aspects of the American petroleum industry which was founded in 1919 after WWI. The API has over 600 member companies, so it is indeed a big player in the US energy industry. Operators in the US OCS are not required to be a member of either the API or the COS.

The basis for the 2010 Workplace Safety Rule (SEMS I) was API RP 75, "Recommended Practices for Development of a Safety and Environmental Management System". COS has since developed the SEMS audit protocol, the requirements for auditors and audit providers and is responsible for the accreditation of the Audit Service Providers. The BSEE SEMS requirements are contained in Title 30 of the Code of Federal Regulations, Part 250 Subpart S (30 CFR 250 Subpart S). When you read this you will see that the CFR incorporates RP 75 and many of the documents and processes developed by COS. The importance of this is that while an operator may not be a member of API or COS, they are still required to comply with 30 CFR 250 Subpart S.

This brings up a question and answer that is generally very confusing to those with Safety Case experience. Who are the Audit Service Providers (ASP) who will be conducting the audit of my SEMS program? Are they part of the BSEE organization? They are not. Are they part of COS? They are not. So who are these ASPs? The ASPs are accredited by COS, and the COS website includes the current list of ASPs. This list is not static and may change for a variety of reasons. The COS is responsible to assure that the ASPs comply with the audit requirements and protocol developed by the COS and incorporated into 30 CFR 250 Subpart S. As an operator you must contract with an ASP to complete the third party audit of your SEMS program as required by BSEE. At this point my friends from the UK get this quizzical look and I know what they are thinking. The answer is yes. The question regards the ability of an operator to competitively bid for the conducting of the audit among the accredited ASPs. This then leads into the concept of the operator being able to select different auditors based upon cost or previous experience with specific ASPs. This usually leads into a long and sometimes spirited discussion of the potential issues with this system. For this book I leave the rest of this discussion to you. My job here is simply to clarify the roles of the players.

Looking at Publication L154, The Offshore Installations (Offshore Safety Directive) (Safety Case etc.) Regulation 2015 (SCR 2015) of July 9, 2015 I found a 164 page document detailing the current Safety Case requirements for operators in the UK "external waters". This is essentially the UK OCS. By contrast, the U.S. Code of Federal Regulations (CFR) 250 Subpart S, "Safety and Environmental Management Systems (SEMS)" is roughly 12 pages. API RP 75, which is included by reference in CFR 250 Subpart S, is roughly 40 pages. There is a lot of duplication between these documents so there is likely less than 50 pages of documented direction for the implementation of your SEMS program. To be fair, the COS has an audit protocol and reference material that supplement these documents, but if a person is used to the granularity in the Safety Case documentation they will find themselves looking for SEMS documented details that do not exist. Don't spend time and resources looking for more; that is all there is.

My Safety Case friends immediately head off into an angry, confused and sometimes depressed state. Take a deep breath and back away from the documents. While the documentation appears lacking when compared with what they are accustomed to, that does not mean the U.S. OCS operations are doomed to operating in some inferior mode. In fact, your Safety Case experience is a big asset when working within the SEMS system. Do not be looking for references to Safety Case, verification schemes, the Health and Safety Executive, the Department of Energy and Climate Change, or the Offshore Safety Directive Regulator. Instead, remember the goal of both the Safety Case and SEMS programs is to identify and mitigate risk thereby preventing prevent future incidents, reduce injuries, and reduce releases to the environment. It is that simple and that complicated.

You have a choice at this point. You can try to revise the SEMS program you are working under to look more like the requirements of the Safety Case. Keep in mind that as you espouse the features of the Safety Case, your audience likely includes people who have no idea what a Safety Case is, nor do they really care. Remember the relative immaturity of the SEMS programs. Hopefully most of your audience has seen or heard of the SEMS Program, and some have experienced multiple audit interviews related to the SEMS program. Understanding this new Safety Case thing is just not likely to be high on their priority list. Instead, become knowledgeable in the SEMS program, and offer your expertize and experience in the language and format of SEMS. For example, identification of safety and environmental critical equipment is critical to both the Safety Case and SEMS. Having someone who has multiple years of experience in the identification and maintenance of critical equipment as part of a Safety Case can be a great advantage to a team in the process of refinement of their SEMS program critical equipment identification process.

My thoughts on this were largely a result of working some very detailed and complex SEMS audits with an auditor who had many years of working within the Safety Case system. I observed him asking questions that while very applicable to SEMS were not the subject of any of the specific documentation associated with SEMS. While on an offshore facility, I was impressed by the efficiency with which he observed and evaluated the operations. Some of this was just the fact that he was a very good and highly experienced auditor. However, in some of the discussions among the ASP team members it was very obvious that his knowledge of the more mature Safety Case system facilitated him thinking at an impressive level of granularity and detail.

Finally, be as comfortable as you can with the relative vagueness of the SEMS documentation. As discussed above, the goal of SEMS is clear. Additionally, SEMS is at its core a Management System (Chapter 3, "Management System Basics"), the concept of which is known to most operations leadership and technical personnel.

Suggested reading

Center for Offshore Safety Website, https://www.centerforoffshoresafety.org/.

Code of Federal Regulations, 30 CFR 250, Oil and Gas and Sulfur Operations in the Outer Continental Shelf, Subpart S, Safety and Environmental Management Systems (SEMS) (7-1-13 edition).

Publication L154, The Offshore Installations (Offshore Safety Directive) (Safety Case etc.) Regulation 2015 (SCR 2015) of July 9, 2015, issued by the Health and Safety Executive, National Archives, KEW, London.

Recommended Practices for Development of a Safety and Environmental Management Program for OCS Operations and Facilities, American Petroleum Institute Recommended Practice 75, May 2008 edition.

CHAPTER 11

The future

Contents

11.1 Where are we now?

Any attempt to get into the details of SEMS and the implementation on the US OCS would not be complete without an attempt to look forward into the future and opine a bit on where it is all headed. There are those who will tell you that the OCS operations are safer than they have ever been, and there are those who will tell you that the BSEE SEMS program has failed to improve operations. In my role as an ASP audit team member and a SEMS SME I have heard people from both camps make persuasive arguments for both cases. This is where my operations manager and emergency management experience kicks in. I learned to be cautious when interpreting the information that comes in during the early stages of an incident. I have experienced both good and bad news early on in an incident which proved to be extreme as the situation clarified. I tend to see the effectiveness of SEMS implementation through the same filter.

I think that we are at that point in the implementation of SEMS where the picture is beginning to clarify, similar to when the information regarding an incident starts to become clear and consistent. We are now in the process of completing the third round of SEMS audits and the second round where an ASP was required. That may well be sufficient feedback to get a clear understanding of the status of SEMS implementation to a degree of granularity that can lead to conclusions and recommendations. My crystal ball is not good enough to know what those conclusions will be, but my experience tells me we should begin by looking in the areas of continuous improvement and repeat findings. If the OCS operators are indeed implementing SEMS as BSEE intends for it to be implemented and as any management system should be implemented, there should be evidence of improvement. The adverse audit findings should become more specific and detailed. I have been involved in all three rounds of the SEMS audit process, and I think I detect some changes in the findings. My sample size is my personal experience, so obviously a more robust evaluation across the OCS is needed, but here is what I think I am observing. For those operators who are conscientiously implementing a SEMS program, the programs have progressed from the early stages of simply working to develop a SEMS program and documents that meet the requirements to having a SEMS program where the documents and processes meet the requirements and the focus is on the fine tuning of the implementation. For example, progressing from not having a documented MOC process to having a SEMS compliant MOC process and then to implementation where the documented MOC process is followed completely and consistently for all changes.

Conversely, repeat findings are a possible indication of the less than fully committed implementation of a SEMS program. Once again, based upon the sample size associated with audits I have been involved with, repeat adverse findings are not uncommon. Let's be brutally honest, operators submit a CAP to BSEE associated with each audit, and provide status updates to BSEE. Add to that the fact that there is three years between audits. Having a repeat finding either means the action item takes more than three years to complete, or that the completion of the action item is not a priority. The level of commitment for specific organizations should be easy to identify. For example, if an operator has had repeated, similar adverse findings with respect to mechanical integrity, one has to question if the operator ever intends to fully comply with the SEMS requirements of mechanical integrity.

However, while I believe the third round of audits will provide significant insight, the realist side of me acknowledges that there will likely never be enough information for everybody to agree on the effectiveness of SEMS implementation. Some will look at the time that has passed since the Deepwater Horizon incident and see this as an indication of success. Others will point out that a lack of incidents has not proven to be a good predictor of the potential for future incidents. The results of improvement in identification and mitigation of risk can look the same as simply being lucky in the time frames we are talking about (approximately ten years). This puts the offshore oil and gas industry in the US OCS in a precarious position. The only sure way to know who is right is to do nothing and wait. If we assume that everything is being implemented and that all operators are progressing through the improvement of their SEMS program, it follows that continuing on the current course should result in a minimized risk of future incidents. On the other hand, if the risks are still in play, and the lack of another incident is due simply to the risks not having aligned themselves such that an incident occurs, then we are taking a huge chance. I think the stakes are too high to not focus on the next level of improvement and performance.

11.2 What will be future keys to success?

Is there a limit to just how far the basic implementation of a management system can take us? I tend to think that there is a limit and base this on my experience in implementing safety programs over the course of my career. When I began my career in the late 1970s (go ahead, do the math), there was a certain acceptance that the oil and gas industry was dangerous and sometimes people got hurt. Nobody liked it, but there was no talk of "target zero" or "perfect days". And to top it off, many operators did not even track contractor injuries only employee injuries. Things changed in the 1980s and into the early 1990s. Corporations began to see that injuries could be prevented and the negative impacts of injuries reduced. So, they hired safety professionals who developed safety procedures to be followed and provided on site safety guidance as available. Many of the safety professionals I encountered during this time period came from other industries or from first responder organizations.

The late 1990s and early 2000s brought further change in that operators began to see that safety was the responsibility of all employees, not just the safety professionals. There is a limit to just how many places the safety professionals can be at one time or in one day. This era saw all employees becoming involved in developing the safety program not just the safety staff. With this came more safety procedures. With more people involved, safety procedures could be developed for all aspects of the operations. It seemed like the answer to all incident investigations was to produce another safety procedure for personnel to follow. I remember volumes of safety procedures. Somewhere in there, it became apparent that an organization could not possibly develop a procedure for every potential safety issue the personnel would face. Enter the era of training people and then expecting them to think about safety, assess their risks and to work in a safe manner. We began to have safety moments in all meetings and investigate all incidents with a focus on prevention. We began to truly utilize "tool box" talks and JSAs.

In January of 2017 I was invited to participate in a workshop sponsored by the National Academy of Science, Engineering and Medicine (NASEM) with the topic of "The Human Factors of Process Safety and Worker Empowerment in the Offshore Oil Industry". Over the course of the two day workshop, I became convinced that the next step change in safety performance will require truly empowered workers. I think the same holds true of the next step in the effectiveness of SEMS implementation.

What does it mean to be an empowered worker? After two days of discussions I decided that it is easy to describe, but so very hard to create. An empowered worker has the training, knowledge and experience required to make decisions regarding work processes in real time and has been empowered by the organization to do so. This is stop work authority on steroids. Stopping the job is the first step, with the evaluation of the situation and the determination of how to go forward being the subsequent steps. I know, this is a bit fuzzy and we operations type like rules and boundaries. I think back over 40 years of how many times a problem that was being evaluated across many levels and technical specialties in the organization was eventually solved with simple elegance using a suggestion from an operations person on the "wrench end" of the work who was comfortable speaking up. Maybe this was a sneak peek at what an empowered workforce looks like?

How will we get there? I don't know but there are a lot of people smarter than me working on this. It is going to involve behavioral and social sciences, human factors research, hazard recognition study, and the science of decision making methods just to name a few. How long will it take? I don't know that either. This is a significant step as it requires, among many other things, a competent and trained workforce with access to all the information needed with which to utilize developed decision making skills. What I do feel confident of is that the next big change in HSSE performance in the offshore oil and gas industry will be shaped and directed by the progress towards a truly empowered workforce.

Suggested readings

Code of Federal Regulations, 30 CFR 250, Oil and Gas and Sulfur Operations in the Outer Continental Shelf, Subpart S, Safety and Environmental Management Systems (SEMS) (7-1-13 edition).

National Academies of Sciences, Engineering, and Medicine, 2018. The Human Factors of Process Safety and Worker Empowerment in the Offshore Oil Industry: Proceedings of a Workshop. The National Academies Press, Washington, DC. https://doi.org/10.17226/25047.

Glossary of selected terms

The following definitions are specific to this book and for use only within the context of this book. These definitions are not intended for any use other than to enhance the understanding of this book. The author makes no representation or warranties with respect to the accuracy or completeness of the contents of this Glossary.

30 CFR 250 Subpart S	Code of Federal Regulations, Title 30 Mineral Resources, Volume 2, Part 250 Oil and Gas and Sulfur Operations in the Outer Continental Shelf, Subpart S, Safety and Environmental Management Systems (SEMS).
AB	Accreditation Body (AB) means a BSEE-approved independent third party organization that assesses and accredits ASPs.
AIChE	American Institute of Chemical Engineers
ANSI	American National Standards Institute
API	American Petroleum Institute
ASME	American Society of Mechanical Engineers
ASP	Audit Service Provider, an independent third-party organization that demonstrates competence to conduct SEMS audits in accordance with the requirements 30 CFR 250 Subpart S
BOEMRE	Bureau of Ocean Energy Management, Regulation and Enforcement. Replaced the MMS in 2010 but was divided into the ONNR, BOEM, and BSEE in 2011
BOEM	Bureau of Ocean Energy Management, manages development of U.S. Outer Continental Shelf energy and mineral resources in an environmentally and economically responsible way. BOEM is responsible for the OCS leasing program.
BSSE	Bureau of Safety and Environmental Enforcement, promotes safety, protects the environment, and conserves resources offshore through vigorous regulatory oversight and enforcement.
CAP	Corrective Action Plan, a scheduled plan to correct deficiencies identified during an audit and that is developed by an operator following the issuance of an audit report.
COS	Center for Offshore Safety, in response to recommendations included in the Presidential Oil Spill Commission investigation into the Deepwater Horizon incident (published January 2011), the American Petroleum Institute (API) approved the charter for the COS in March 2011.
COW	Control of Work, management systems used to ensure that work is done safely and efficiently
CPS	Center for Chemical Process Safety
CSB	U.S. Chemical Safety and Hazard Investigation Board

EMS	Environmental Management System
EPP	Emergency Preparedness Plan
GOM	Gulf of Mexico
HA	Hazard Assessment
HAZID	Hazard Identification, a hazard assessment method that is a qualitative technique for the early identification of potential risks
HAZOP	A Hazard and Operability study, a structured and systematic examination of a complex planned or existing process or operation in order to identify and evaluate problems that may represent risks to personnel or equipment.
HSSE	Health, Safety, Security and Environment
Human Factors	Human Factors is the study of how humans behave physically and psychologically in relation to particular environments, products, or services.
ISO	International Organization for Standardization
IT	Information Technology
JSA	Job Safety Analysis
LOTO	Lock Out Tag Out
MMS	Minerals Management Service was an agency of the United States Department of the Interior that managed the nation's natural gas, oil and other mineral resources until 2010
MOC	Management of Change
MODU	Mobile Offshore Drilling Unit
MOM	Marine Operations Manual
NACE	National Association of Corrosion Engineers
NDT	Non Destructive Testing is a group of analysis techniques used in industry to evaluate the properties of a material, component or system without causing damage.
OCS	Outer Continental Shelf, the submerged lands, subsoil, and seabed, lying between the seaward extent of the States' jurisdiction and the seaward extent of Federal jurisdiction.
OIM	Offshore Installation Manager, the most senior manager of a floating offshore platform. This position has requirements set by the USCG.
OMS	Operations Management System, the management of systems or process responsible for managing the core processes, in this book those core processes are those used for oil and gas operations
ONNR	Office of Natural Resources Revenues, manages and ensures full payment of revenues owed for the development of the Nation's energy and natural resources
OSRP	Oil spill Response Plan, a planning document prepared and used by industry owners and operators to respond to a worst-case discharge from their offshore facilities. Requirements are outlined in 30 CFR Part 254
PIC	Person in Charge, means the person on each facility to whom all personnel are responsible. For purposes of this book the most senior manager of a fixed offshore platform
PPE	Personal Protective Equipment
PSM	Process Safety Management
PSR	Pre Start - Up Review

ROV	Remotely Operated Vehicle
RP 75	American Petroleum Institute Recommended Practice No. 75, Recommended Practices for Development of a Safety and Environmental Management Program for OCS Operations and Facilities, May 2008 Edition
S&EI	Safety and Environmental Information, Element 2 in the SEMS program
SDS	Safety Data Sheets, a standardized document that contains occupational safety and health data for chemicals
SEMP	Safety and Environmental Management Plan, essentially the same as a Safety and Environmental Management System.
SEMS	Safety and Environmental Management System, for the US OCS this generally refers to the requirements of 20 CFR 250 Subpart S
SIMOPS	Simultaneous Operations
SWA	Stop Work Authority, Element 14 in the SEMS program
SWP	Safe work Practices
USCG	United States Coast Guard
UWA	Ultimate Work Authority, Element 15 in the SEMS program

Index

Printed in the United States
By Bookmasters